P9-CED-733

Consumption Economics

The New Rules of Tech

J.B. Wood
Todd Hewlin
Thomas Lah

ISBN: 978-0-9842130-3-0

LCCN: 2011917022

Printed in the United States of America.

Table of Contents

Foreword

By Geoffrey Moore

Author of *Crossing the Chasm, Inside the Tornado, Dealing with Darwin,* and *Escape Velocity*

As business categories evolve and mature, they tend to bifurcate into two business architectures: one specializing in *complex systems,* the other in *volume operations.* Prior to this split, only the complex systems model thrives, focusing on institutional customers, syndicating responsibility for the system across an ecosystem of vendors, partners, and the customers themselves. Because of the cost and complexity of these demands, there is no place for consumers or small businesses in this model, and even midsize businesses can find themselves challenged.

This has been the case for enterprise-class IT systems throughout the course of my life. But as J.B., Todd, and Thomas make clear in this book, it will no longer be the case going forward. The rise of the volume operations IT model, initially confined to simplified consumer transactions, has now breached the levees protecting the complex systems enclaves. Riding a wave of "______ as a service" business models that stretch from infrastructure as a service to software to platforms to business processes, the volume model is reengineering the landscape—and the profit pools—of enterprise IT worldwide.

Now let me be clear. It is not that the complexity of enterprise IT has gone away. Indeed, if anything, it has increased. But as this book shows, responsibility for managing this complexity is shifting from the customer to the vendor. This has huge implications for vendors, for IT organizations inside the enterprise, and for the business itself in terms of leveraging IT, be it for differentiation, expansion, or productivity.

On the vendor side, the shift from a system to a service orientation fundamentally reshapes the economics of the IT sector, migrating zones of profitability to new roles and new players. Systems vendors, long accustomed to being top dog, now find themselves herded into utility computing platforms that disintermediate their relationships with the end customers. Conversely, customer service teams, traditionally marginalized as cost centers, now have taken center stage as their transactions have become the core of the next-generation offer set.

This direct interaction between the system and the end user enables next-generation IT vendors, in turn, to disintermediate the IT organization, at least in its traditional role as gatekeeper. This is not to say that enterprises no longer need an IT function, but to keep up with the times, IT must shift its focus from building and maintaining systems to shaping the interface between managed services providers and the business processes of their enterprise. The impact of staffing should not be underestimated, as the skill sets required to support the new model are quite different from ones used in the old model.

Finally, as vendors and IT organizations are responding to this massively disruptive turn of events, business executives and department managers also must come to the table with a new mindset. "IT as a service" dramatically lowers the barrier to entry for every enterprise. If yours does not take advantage of this new freedom, you will find yourself falling behind your competitors who do. Never has it been easier to gain a return from IT spending, and it is critical that organizations leverage this newly released cornucopia of capabilities smartly in order to keep current with an increasingly challenging global economy.

But how do you actually do that? How do vendors and IT organizations and business teams co-evolve to capitalize on this dramatic change? That is the focus of this book. Its authors have worked at the epicenter of this disruption throughout the past decade, and I can think of no one more qualified to frame the new agenda.

So buckle up, for it is bound to be a bumpy ride, and learn from these pros how to negotiate the hairpin turns in your future.

Geoffrey Moore

1 How Good We Had It:

The Money-Making Machine Known as High-Tech

In sector after sector, the profitability of selling tech products is shrinking. The total amount of gross profit dollars in some tech product sectors is getting smaller every year. For the first time, most companies no longer seem able to innovate their way around the problem. Price competition is taking a huge toll on product margins not just in consumer sectors like PCs, but also in traditionally profitable enterprise segments like networking and software. It's not because tech companies can't innovate anymore, but because tech customers can barely use the complex technology they already own. They have unused features and licenses, excess capacity, and stable systems that serve their basic needs, so why buy more unless it's cheaper? This is not the path to profitable growth. It is the path to commoditization, and it is forcing companies to change what business they're in just to maintain margins. Hardware companies are jumping to services and software. Software companies are making their profits off of maintenance contracts, not selling software.

As an industry, we face a growing gap between what our products are capable of doing and what our customers actually do with them. This may be the true gating factor in the growth of tech industry profits and it's not a problem that more features will cure. Failure to close this gap means future revenue growth that is anemic, unprofitable, or both. In the new economics of the cloud, driving usage is even more critical since nearly all the revenue will be based on consumption, and switching costs will be low for disgruntled customers. The cloud as it is currently envisioned is a tricky thing. On the one hand, it could be the rapid draining of the profit pool for high-tech—a relentless drive for cheaper and cheaper versions of standard functionality. At the Technology Services Industry Association (TSIA), we believe it could also enable a unique ability to fight commoditization. This is what this book is about. However, before we can look to the future, we need to take a look at how we got where we are today.

Technology purchase decisions have always been based on an assessment of risk and reward. If the reward was truly compelling, customers would take the risks—pay a high asking price, pay it all up front, put up with lots of frustrations, lots of complexity, even risk failure. Why would a buyer enter into such a lopsided agreement? It's simple: Tech offers were *that* compelling. Not just compelling, but *Wow!*-compelling. After all, *Wow!*-compelling offers are what high-tech companies specialize in. It's what VCs look for, what boards of directors and financial analysts reward, what R&D folks live for, and what salespeople want to sell.

For decades our offers have compelled corporations to buy complex and sophisticated hardware, software, networks, and services. While the investment risks to the buyers were significant, they gladly placed their bets because of the huge opportunity for value creation. Successful IT systems promised to dramatically increase revenues, cut costs, create competitive advantage, improve customer service, or boost employee productivity. These were rewards worth taking real risk for, even if it meant that the customer shoulders far more of it than the tech seller does.

As a consumer we experience the same phenomenon. We make trade-offs too. Hundreds of dollars and lots of personal time are required to realize the potential reward of some compelling technology offer. But instead of improving corporate profits, we just want to edit digital photographs or to entertain ourselves with awesome music and games. So no matter if you are a CIO spending $5 million on CRM or a mom spending $500 on little Bobby's new laptop, almost everyone has stared down the barrel of the technology risk/reward decision. In either case, as long as the tech company could innovate in dramatic fashion, they got the customer to pay handsomely up front and assume the risk.

So how has this worked out? Pretty darn well for the tech companies. There has been so much breakthrough opportunity that the right product could catapult itself, its company, and its users to stardom. If you were the tech company with a hot product, you were golden. As a technology company, your goal was to join the famous parade of multibillion-dollar product franchises like Cisco routers, Microsoft Windows, Oracle databases, HP printers, VMware virtualization software, and the Apple iPhone. The list of highly profitable hardware and software products is long and impressive. Each deserves its place. These products created revolutionary opportunities for customer value. Even though the hot product rarely delivered on all cylinders, it usually worked well enough to get exciting results.

In addition to fast-growing revenues, successful technology companies became associated with *very* successful business models. These models were well documented in *Bridging the Services Chasm*, a book we published a few years ago. Our analysis of extensive industry benchmark data allowed us to model how successful technology companies generated their margin dollars. What we have not discussed in any of our previous works is the underlying purchase pattern that fuels all technology business models. In enterprise tech markets it looks like this:

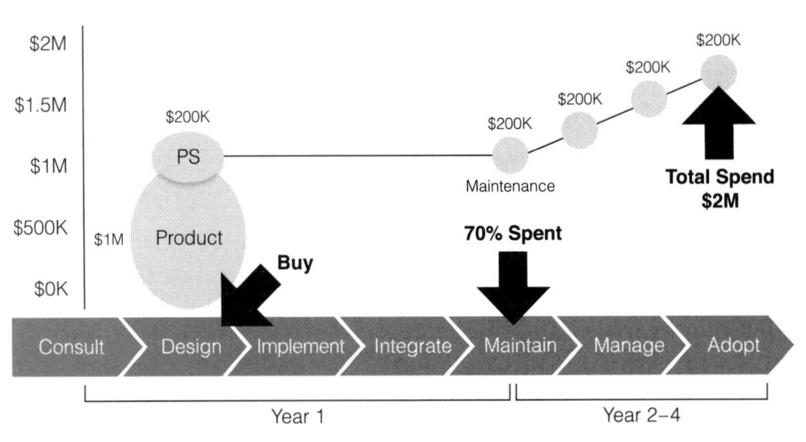

FIGURE 1.1 Typical Corporate Technology Purchase Pattern

The typical corporate IT buyer went through a selection process and picked the product(s) they wanted. They paid upfront for the product and for the installation, integration, training, and other professional services needed to get the product operational. Since the system would likely have a useful life of several years before major components would be replaced by next-generation offers, they wanted the software updates along the way and insurance against downtime. Thus, they also agreed to buy coverage for those products through an annual maintenance agreement offered by the supplier that they would continue to renew during the useful life of the solution.

What is really important about this picture are not the dollar values. In this example, the total cost of the system and related services is $2 million over a four-year period. Plug in the relevant number for one of your customer's typical purchases. It could be $20 million or it could be $2,000, no matter. The key point is that by the time the products are actually getting adopted and used, at least 70 percent of the total four-year investment has already been spent. That money is now in the tech company's bank account. And if the customer purchased maintenance, which 80 to 100 percent of corporate customers do, the probability that the tech company will get the last 30 percent of the money is nearly guaranteed. At "go-live," the customer has huge sunk costs in the project. The idea of switching to another supplier is almost laughable. That customer is locked in.

Tech companies have spent decades optimizing their product development, marketing, sales, and services around this purchasing model. It also represents the financial model that Wall Street understands and can value.

Not necessarily unfair, right? Big commitments are what customers make when they buy important products. They buy cars and houses up front, so why not technology? After all, think about the investment that the tech company had to go through to get that deal:

- The development costs to build the products.
- The cost of any included hardware, third-party software, and licenses.
- The marketing costs to promote the offer.
- The sales salaries and commissions to get customers to buy it.
- The pre-sales consultants that support the sales process.
- The service personnel to implement the system and train the users.

Add that all up, and the money spent by the tech company to win that one piece of business is substantial. But they could relax because they were getting a big up-front check from the customer. The revenue associated with the initial product sale not only covered all those direct and indirect deal costs, but often returned handsome profit margins to the tech company. So who shouldered the risk in these scenarios, the customer or the tech company? Well, as we just pointed out, the tech company had plenty of up-front costs, but they got paid for those costs (and more) in the first year. So while they may have some temporary issues related to the timing of their expenses vs. their revenue recognition, billing cycle, and cash collection, their absolute risk was minimal.

What about the customer? Well, they were the proud owner of some bright, shiny, new technology. They took the risk on its potential and wrote the big check. So where does their reward come from? Let's be honest, their reward is one hundred percent dependent on actually deriving the benefit from the new system that they were promised by the seller. The system has to actually work and has to be consumed successfully. The business users have to change their workflow from the "old way" to the "new way." So who has primary responsibility for making sure all that happens? Customers do. It is a fact that, in nearly every technology product market, customers assumed a disproportionate share of risk in the risk/reward purchase decision. They were the party responsible for making the decision pay off, and to do that they faced a big problem.

Back in 2009, we published the book *Complexity Avalanche*. In it we documented a phenomenon that is now generally recognized as gospel truth in tech. That phenomenon is called the "Consumption Gap."

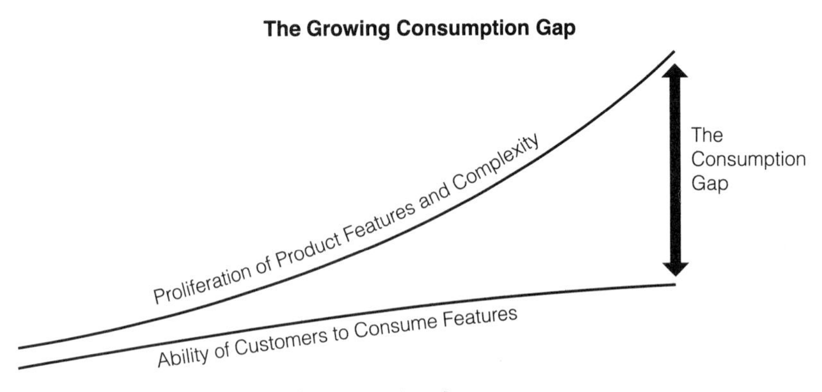

FIGURE 1.2 The Growing Consumption Gap

The Consumption Gap is based on the idea that technology companies can add features and complexity to their products at a much more rapid rate than their customers have the ability to consume them. The result is a growing gap between the potential value that products and services can provide vs. the business value that customers are actually achieving. The proof litters the industry. Software upgrade rates are slowing, the percent of features in use are declining, and product sales cycles are lengthening as customers struggle to digest all the product features they already own. The end result is that technology refresh rates in many product categories are slowing as "good-enough IT" is becoming a reality. Almost every company is feeling it . . . bet yours is too!

Those trends are how the Consumption Gap negatively affects things from the perspective of the tech company. But how does it present itself from the customer's side? Consider these statistics:

- The average effective usage rate of enterprise software is only 54 percent.[1]
- Only 14 percent of enterprise software deployments are rated as "very successful" by the IT execs who own them.[2]
- Of the nearly $14 billion in consumer electronics that were returned by customers to retailers in the U.S. during 2008, only five percent were actually broken.[3]
- Hard-to-use software is behind the leakage of sensitive health data online, according to a study by Dartmouth researchers. "So many of the systems, particularly in health care, are bad," said the study's co-author, M. Eric Johnson. "So people find workarounds—using word-processing and spreadsheet tools that aren't secure."[4]

The Consumption Gap is huge and its impact on the tech economy, and society in general, is almost incalculable. Sure, there are ranges in the severity of the Consumption Gap—the more software you have in your product, the more likely you will have a gap. But name a product category that isn't rapidly adding capability and complexity through software? As products evolve, the gap between what products can do vs. what they actually deliver is getting larger, not smaller. Forums on Microsoft's Word product estimate that there are now over 1,200 features available in this product.[5] What percentage of those features is being used by the typical Microsoft Word user? A TSIA member company just released the latest version of their software and they added 1,500 new features. Fifteen hundred on top of what was already there. Really?

Complexity Avalanche laid out many key tactics that tech companies could adopt to stem the Consumption Gap by inserting themselves actively into the process of monitoring and driving successful customer outcomes. It showed them how they could do this by repurposing and restructuring their existing customer service and support assets. Some companies like Xerox, Apple, and GE Healthcare have made the decision to take on that market-leading role. They realize that getting customers on a steady path of capturing more value from their existing technology is not only a competitive advantage—it is the key to making them hungry for more product capability.

Why has the industry gotten away with the Consumption Gap for so long? Once again, it has been a function of the *Wow!*-compelling nature of our offers and the purchasing model for customers. The potential rewards that our offers proposed were so compelling that customers accepted the shortcomings, challenges, and risks. Once customers made the purchase decision, they were locked in. There was no turning back. They had to learn to work around the frequent realities of:

- High implementation costs (sometimes five times or more than the product cost!).
- Lack of interoperability.

- Extraordinary complexity.
- Frustrated business leaders.
- High cost of ownership.
- High switching costs.
- Poor end-user adoption.
- Hard-to-measure business benefits.

We knew it. Our customers knew it. But our offers were *that* compelling. No one wants to say this, but tech companies have been guilty of some over-promising and under-delivering for decades. Our customers lived with that because they were committed, and they had to drive through whatever obstacles they encountered to get to the *Wow!*-compelling rewards. Most customers did so to one degree or another.

Here is the bottom line: If you were the tech company, the questions of how well your products got used or how often did not usually have a meaningful, immediate financial impact. It was, after all, a product model, and we got paid up front. Of course, no tech company could survive repeated customer failure. That is not what we are talking about. We are talking about under-delivering on the full potential of the product. Tech companies learned that "good enough" consumption of value was all they needed to achieve for their economics to be optimized in this model.

These companies do care about customer satisfaction. The authors of this book have been working with technology companies for years on tactics to improve customer success through well-executed service experiences. But let's be honest, profit-driven companies have to prioritize profits. And for technology companies, on any given deal, profitable revenue was locked in before any end-user consumption began. The product revenue, the professional services revenue, and the first year of maintenance were in the bank, whether the customer achieved the project return on investment on the technology or not. It's not like we, as a technology industry, didn't care about the customer's successful consumption of value. We did. We knew that badly under-performing customers would be harder to up-sell later and may not act as positive references for our products. In the case of very important customers, we would spend additional time and money to rescue the situation. But the majority of customer Consumption Gaps went unnoticed or unresolved by the tech company. Solving them would add to this quarter's costs, but not its revenue. So the company's urgency to attack the problem was somehow less immediate, less real, than hitting "the number." Rather than fix many of the problems that we knew were contributing to the Consumption Gap, we focused everyone on making this quarter's revenue and cutting this quarter's costs. So we did what we could afford to do according to our "product playbook," and that was about it.

This business model has worked. The tech industry has been a money-making machine. Running this play from the product playbook has created trillions of dollars in shareholder value and millions (in a few cases billions!) in wealth for thousands of industry executives.

But will it continue to work?

Up until recently it sure seemed that way. In the middle of the last decade, tech had recovered from the 2001 dot-com fiasco and was starting to rock once again. Things looked great, and by 2007 we were again running the product playbook furiously and successfully. But if you looked closely, if you talked to frustrated corporate CIOs, the handwriting was already on the wall. There was a steady, growing pile of customer frustration—dead wood waiting for a match. Big changes were going to come to the tech industry, specifically enterprise tech, as maturity set in. And sure enough, around 2008, the match appeared.

But no one saw it coming because it was an unusual-looking match.

Three things within fifteen months—not directly related but highly synergistic—occurred that will forever change the rules of the tech business.

1. The global economy tanked.
2. Cloud computing got hot.
3. The iPhone came out.

Let's start with the economy, because it threw gasoline on the fire of the other two.

At TSIA we have a piece of nostalgia, we believe. It's a copy of the *McKinsey Quarterly* from Q3 2008. That is exactly when the meltdown occurred. Exactly. Take a guess what amount of the editorial content in the Q3 2008 *McKinsey Quarterly* was devoted to the impending meltdown? Zero. AIG? Citicorp? Housing bubble? The impending global meltdown of financial markets and credit? Zero. Every article was about the growing influence of China. Now wait a minute! These are the smartest guys in the world, right? How could they miss this? Well that's because everyone missed this: Alan Greenspan, the IMF, the SEC, Jim Cramer, and your local Rotary Club.

Then it hit. And it would have profound short- and long-term consequences on tech.

We have already lived through the short-term ones. Tech customers large and small stopped having confidence in their top line. Revenue levels became very, very uncertain. Smart corporate executives did the logical thing—they hunkered down and took a red pen to every discretionary and many nondiscretionary costs. Not just a little focus on expense reduction but a maniacal one. "How can we innovate our cost structure?" they wondered. They cut this, slashed that, rethought this model, and outsourced that one. It was beautiful execution. It could be argued that this was one of the best performances in the history of business management. Corporations protected themselves surprisingly well by rapidly rightsizing costs to fit their lower revenue outlook. Thanks to smart management and some innovative government intervention, we righted our sinking ships. It was a great piece of work.

There were also a couple long-term lessons seeded in that rough experience.

One is that corporations could still function effectively with far fewer staff than they had before. Yes, there were gut-wrenching decisions required to make the cuts. And, yes, unemployment in the developed nations was the result. But you know what they learned? They were still alive, still functional, and still able to conduct business. Almost minimally impaired! No wonder we have a jobless recovery. While it is bad in that respect, it is good for tech because these customers are already preferring tech-driven productivity improvements over staff additions as their businesses pick back up.

Another lesson is that companies could successfully pressure their suppliers to do more for them. In the case of the tech industry, that meant customers putting off new projects, negotiating maintenance prices down, extending the life of existing equipment, and deferring software upgrades. The kicker is that they also started demanding that suppliers put "skin in the game" to drive the value of the last round of purchases. A tipping point had been reached. CIOs, many with less IT staff as a result of cost cutting, still had a mandate to eke more business productivity and value from their existing IT stack. One repercussion was an important refinement in the discussion they had with their tech suppliers around what "do more" actually meant. To be specific, four new thoughts got into the mix:

1. IT execs hit on the idea that they could lower their costs by simply lowering their complexity.
2. The notion of "good-enough IT" took hold. Upgrading and replacing shifted from a habit to a heavily scrutinized decision.

3. Customers started talking about sharing risk—tying their tech spending to business value realization and shifting from up-front payments to pay-as-you-consume (or even pay-on-business-value-capture).
4. And finally, they demanded that the supplier's own people get involved in their production processes to supplement or replace internal staff.

Basically they were asking for better alignment between their suppliers' remuneration models and their internal business case. In just two dazzling years between 2008 and 2010, the maturing of enterprise tech went from a walking pace to a run—more price pressure and more accountability. Business results began to trump cool features, and the cost-saving power of simplicity became evident. Just as customers figured out that they could get by with less staff, they also figured out they could demand, and get, different behaviors from their tech suppliers. They lost some interest in the high-priced, high-risk, high-complexity offers that created trillions of dollars of wealth for high-flying, Silicon Valley-type companies.

In short, by 2010, tech execs were being told by their biggest customers that the old product playbook was beginning to go out of style. It is against this demand for a new tech economic model that the two other game-changing tech trends got fueled.

Cloud computing was a model that, at least conceptually, was perfectly suited to these new kinds of customer demands. It was, by design, pay-for-consumption. It promised to be simpler and faster to implement, plus you only paid for what you used. Why take on the complexity when your supplier can deal with it? Why assume risk in the old pay-up-front capital expenditure (CapEx) model when you can align revenue to expense in a pay-as-you-go operating expense (OpEx) model? The attraction of these value propositions got traction fast. Today cloud is not just an IT trend; it's a legit competitor on almost every new IT deal. For this same reason, there is also growing interest in other OpEx models like managed services.

The final disruptive force was the arrival of Apple's iPhone and App Store. Originally viewed as a consumer phenomenon, it did not take long before the industry began to wake up to its implications on the corporate computing models of the future. The balance of power in corporate IT had always rested squarely on the IT department. They made the decisions, mandated the usage, and controlled the environment. Not very Apple-like thoughts, and not how the iPhone worked. Apple encouraged its end users to make their own decisions, to build "their own" device. There is a reason it's called the iPhone and not the ITPhone. The individual iPhone users chose the software they wanted, content they liked, and services they subscribed to. They had thousands of low-cost app and content options from which to build a device that truly suited how they lived, played, worked, and communicated. And you know what? iPhone users felt much more productive in their personal lives than they did in their professional lives. They were more up-to-date on what was happening each day with their friends and family than they were with their team member three cubicles away. The existence of 99-cent apps, an apps marketplace, user-defined capability, low-cost content, work/life blended devices . . . these are breakthroughs that are simply not going to stay out of the enterprise. People did what worked for them, and that plan worked better than being told what to do by some IT department policy. The consumerization of enterprise tech began to knock on the IT department's door.

Put these three things together—increased scrutiny on technology ownership costs and ROI, the disruptive promise of cloud-computing models, and the consumerization of IT—and you have the underpinnings of a sea change in how tech companies must operate to drive profitable growth. A brutal and uncertain economy has created a permanent demand for simpler, lower-cost, *lower-risk* technology solutions. The successful supplier of the future will have to shoulder much more, or perhaps all, of the financial risk of customers taking on their products and services. They will have to invest mightily in the customer's success. Not just to set up the products, but also to help the customer's end users actually consume their capabilities—sometimes even doing it for them. And the power shift from the IT department to the business users, maybe even the actual end users, will take a huge leap forward. Once off the radar for most enterprise tech companies, the end users are about to emerge as *the* key players in the next phase of the industry's development.

A shift is beginning. In the growing worlds of cloud and managed services, the pattern of purchasing technology is changing. With this new purchasing pattern, risk is moving from the customer to the tech company.

So regardless of whether your company is facing declining margins in your traditional product markets or starting to understand the profound risk-shifting impact of the cloud, the rules of the game are changing. The old play from the product playbook has run its course. The ability to successfully drive consumption is about to become the critical enabler of profitable growth. How we build products, how we drive revenue, what services we offer, what we must do to succeed are all on the table. It is the rise of Consumption Economics, and it has new rules. Here is how we think they might play out . . .

2 Shifting Clouds and Changing Rules

THE CLOUD HAS BUZZ. THE MEDIA IS CITING IT AS THE NEXT "BIG thing" in tech. Traditional tech companies are running around figuring out who to acquire to stake their claim in the next big gold rush. But wait! What, exactly, is the big deal?

Larry Ellison, CEO of Oracle, is on record as saying the cloud is just the mainframe reincarnated, and he has a good point.[1] Is it really a big deal where the software and hardware are hosted or how cheap an endpoint device you can get away with? No. That was the whole mainframe model. We would also argue that it's not whether you rent or own. The "______ as a service" model has received a lot of focus. But paying over time rather than paying up front is not unheard of. Capital equipment has been leased or financed for years.

So, then, what is the big deal about the cloud? At TSIA, we think seven shifts will profoundly affect how tech companies operate, differentiate, and make money:

1. The risk in the purchase decision will shift from the customer toward the supplier.
2. Complexity's long and illustrious reign will end. Simplicity will be king.

3. Cloud customer aggregators will shrink the direct market for tech infrastructure providers.
4. Big changes will come to the channel ecosystem.
5. Cheaper enterprise software will emerge.
6. IT departments will "get out of the way" of end users.
7. Tech companies will capitalize on user-level behavioral data.

Shift #1: The risk in the purchase decision will shift from the customer toward the supplier.

Remember the drawing in Chapter One about the typical enterprise IT purchase pattern?

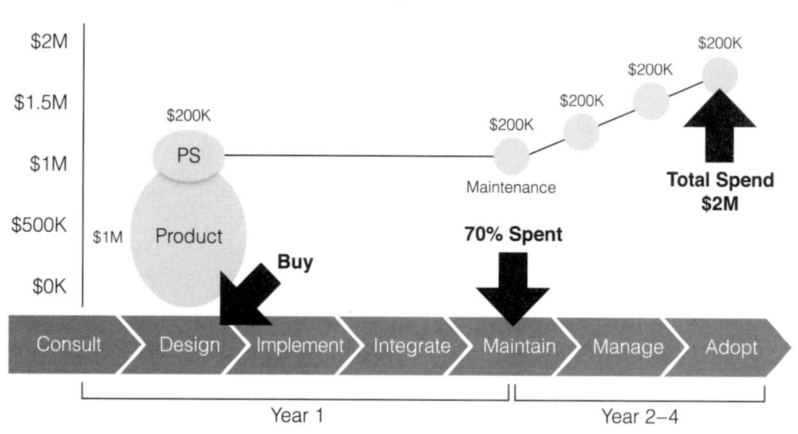

FIGURE 2.1 CapEx Technology Purchase Pattern

Let's call it the "CapEx purchase model," because the customer is paying a price to own an asset. It could be hardware, software, multifunction output devices, or CT scanners. It could be paid in cash, financed, or leased—no matter. The price is set, the deal is struck, and the seller is guaranteed payment. In this model, the tech company only has one stress point: winning the deal. All the risk then rests on the customer.

What will this look like in a cloud model? The answer could be radically different as we shift from an up-front CapEx purchase to a pay-as-you-go OpEx purchase.

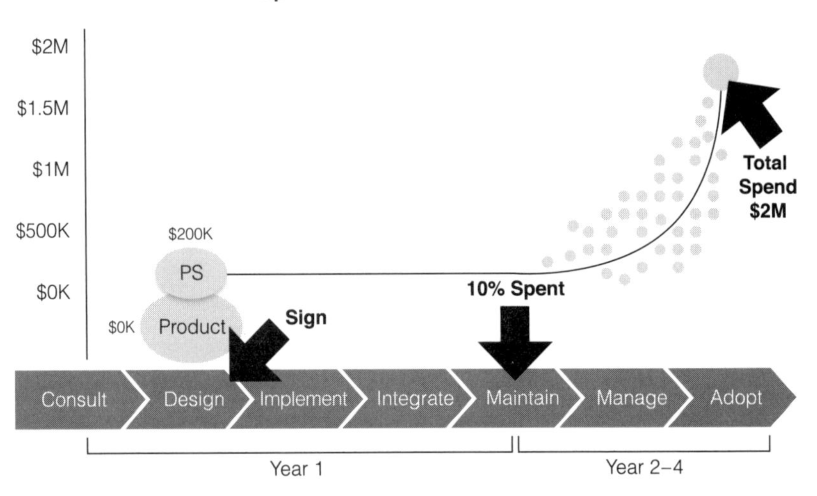

FIGURE 2.2 OpEx Cloud Purchase Pattern

As a supplier, you still have to go through many of the same expensive steps to win a cloud deal. You have to develop the product, do the marketing, execute the sales cycle, bring in the presales consultants, and win the deal. So in this respect, there is not a bit of difference in the cloud model. So where is the difference? It's simple. In the traditional CapEx model, there was a pot of gold waiting at the contract signing. Congratulatory voicemails were sent, pizza was ordered, beer was consumed, and commission checks were cashed. In the cloud, the supplier might incur all those same costs, win the deal, and get absolutely nothing for the trouble. Actually, to be precise, you get the pot. It is empty, but it is *your* pot, and you have earned the chance to fill it with gold from this customer over the years ahead.

There is no big fat circle called the product sale. What you have earned is the right to start filling up your new pot with all those little dots shown in *Figure 2.2*. Even though your sales reps will push hard to get as many minimums, guarantees, and commitments as they can from the customer, at the end of the day, most all the revenue will be in the future.

What are all those little dots? They are billable transactions—new users, added features, content downloads, more modules, extra computing capacity, more storage, more, more, more. And, if consumed by a user, you earn the right to bill for them. That is good. But they really *are* little—tiny, in fact. They are micro-transactions. Let's take a closer look:

Below is the 2011 price list for Amazon's EC2 cloud-based computing services.

Amazon EC2 Pricing

Region: US – N. Virginia

Standard On-Demand Instances	Linux/UNIX Usage	Windows Usage
Small (Default)	$0.085 per hour	$0.12 per hour
Large	$0.34 per hour	$0.48 per hour
Extra Large	$0.68 per hour	$0.96 per hour
Micro On-Demand Instances		
Micro	$0.02 per hour	$0.03 per hour
High-Memory On-Demand Instances		
Extra Large	$0.50 per hour	$0.62 per hour
Double Extra Large	$1.00 per hour	$1.24 per hour
Quadruple Extra Large	$2.00 per hour	$2.48 per hour
High-CPU On-Demand Instances		
Medium	$0.17 per hour	$0.29 per hour
Extra Large	$0.68 per hour	$1.16 per hour
Cluster Compute Instances		
Quadruple Extra Large	$1.60 per hour	N/A*
Cluster GPU Instances		
Quadruple Extra Large	$2.10 per hour	N/A*

* Windows® is not currently available for Cluster Compute or Cluster GPU Instances.

Source: Amazon website, July 2011. © Amazon.com.

FIGURE 2.3 Amazon EC2 Price List

Prices for a Small, Standard On-Demand instance begin at around 10 cents per hour. That's right—10 cents. Now *that* is a small dot. But don't fret; you haven't seen the price for the Cluster GPU Quadruple Extra Large instance. That dot is a whopping $2.10 per hour.

A couple of interesting observations here: The obvious one is the small average unit price for the transactions. Amazon adds up all the dots (transactions) and bills the customer. These dots are a far cry from the fat up-front server circles that IBM, HP, and Sun used to sell. But now IBM and HP are selling dots. In general, the total price direction of the cloud is likely to be down, and we don't mean just from the current CapEx price levels. Standard, published prices like these are also likely to spark classic retail price pressure from competitive cloud service providers. Back in 2006, Jeff Bezos, CEO of Amazon, delivered the opening keynote at MIT World. In this keynote, he overviewed the economies of scale that will allow Amazon to drive down the costs of acquiring computing infrastructure.[2] Clearly, the Amazon model is to succeed on razor-thin margins. Bezos does not fear plunging prices for storage and computing cycles; he wants to be the guy who leads it there!

A second thing to keep in mind is that, even at these low prices, the simplicity of access means there is not much holding the customer to this supplier. If customers don't like subscribing to computing power from Amazon, it is not much of a switch to move to another cloud-based computing and storage supplier. This low switching cost could take the threat of retail price wars and commoditization to tech sectors that have never experienced anything like it before.

Okay, you say, that may be true for commodity examples like this, but what about something more complex? What about enterprise software? Below is the 2011 price list for salesforce.com's core CRM applications.

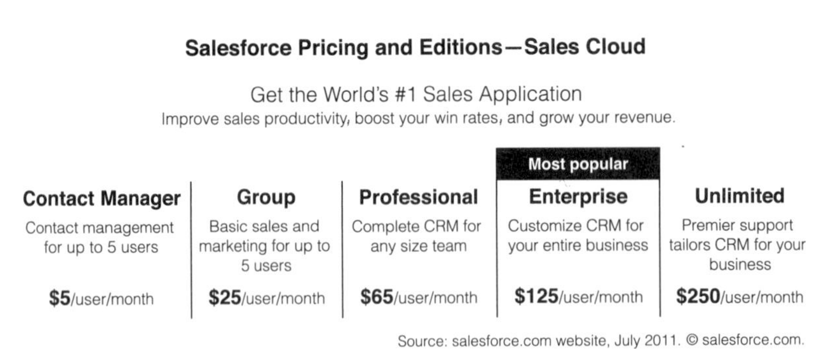

Salesforce Pricing and Editions—Sales Cloud

Get the World's #1 Sales Application

Improve sales productivity, boost your win rates, and grow your revenue.

Contact Manager	Group	Professional	Enterprise (Most popular)	Unlimited
Contact management for up to 5 users	Basic sales and marketing for up to 5 users	Complete CRM for any size team	Customize CRM for your entire business	Premier support tailors CRM for your business
$5/user/month	**$25**/user/month	**$65**/user/month	**$125**/user/month	**$250**/user/month

Source: salesforce.com website, July 2011. © salesforce.com.

FIGURE 2.4 Salesforce.com Core CRM Applications Price List

The basic Contact Manager module can be subscribed to for $5 per user per month—another very small dot. But maybe salesforce.com can get their power users to buy the biggest dot: $250 per month for unlimited functionality. By historical standards, this is still a pretty small dot—still a micro-transaction in the enterprise world. (P.S.: If you were thinking that you could apply the "this is a commodity" argument again here, remember that this is sophisticated CRM, not *Angry Birds*. Just five years ago Oracle paid $6.1 billion to buy Siebel. Do you think that Oracle thought enterprise-worthy CRM would be selling at these prices in 2011? Don't think salesforce.com's product is being used by big enterprises? Think again—Cisco, McKesson, and Qualcomm all use salesforce.com. What's worse, as we will discuss later, is that even these price points are under attack. This show is definitely coming to a theater near you.)

There is a third interesting point that these two examples share. We already know that all the dot (micro-transaction) prices are small, *but* let's not lose sight of the range of values. In the salesforce.com example, the swing in what they can bill for that one user in that one month is 5,000 percent between the most basic feature set at $5 per month and the most advanced, unlimited feature offer at $250 per month. The same is true for the Amazon EC2 offer. The price range as a percentage is huge. We will come back to this in detail later in this book, but the takeaway is this: Even though all the micro-transactions are tiny, both volume and average selling price remain critical components to the billable revenue formula. Successful cloud suppliers need to optimize around a model that systematically and proactively drives both.

The net economic impact of cloud, service-based, subscription-based, managed service, or other OpEx offer models is this: The prices are coming down, and the risk in the deal is shifting. The costs to win the deal are largely unchanged. But rather than a pot of gold on contract signing, all you get is the pot. The risk of whether it eventually gets filled up with the customer's gold is completely on you as the supplier. If the customer doesn't get value, doesn't actually achieve high volumes of usage, or finds a better, cheaper OpEx offer next year from a competitor, who is at risk? You, the supplier, with all the sunk, up-front cost. Even in current leasing models, the customer is on the hook for the whole purchase price, usage or no. The cloud represents the first large-scale model that shifts risk squarely from the customer to the supplier. Remember these simple words: *"No usage, no money."*

There is a famous sales axiom: "The most expensive place to finish is *second*." This axiom referred to a competitive deal where your salesperson made it all the way to the end. You spent all the time, travel, bandwidth, and brainpower needed to get right to the finish line, but then lost out to a competitor on the final decision. In a cloud or a managed service contract, there is a new possibility: The most expensive place to finish may now be *first*. Since the customer's financial commitment is tied primarily to usage, you may not only have to sell, but depending on how you bundle your products and services into your subscription, you may have huge up-front costs just getting the customer ready for the billable consumption that starts the filling of your pot. If the customer largely fails to adopt, you could be more upside down in that deal than the supplier who came in second. As a matter of fact, in a managed services contract, it can be even worse. In addition to all the sales costs, you may also be installing equipment on-site on your own nickel, even putting people on-site to operate some of it—all before you have seen one penny of revenue from the customer.

As we said, every supplier will be trying to get as big an up-front customer commitment as they can. Minimum usage, minimum contract length, etc. are all favorite requests. But what do you think the trend will be over time? We believe that more and more contracts will be tied completely to usage and/or business outcomes. So it is great if you are easing your way today into the big risk shift by demanding contract minimums, but the trend is clear: Risk is shifting in your direction. How long it takes the tech company to make a profit on an individual customer has completely changed. Instead of being profitable from day one, it might be year three. The front-end costs are now on the tech company's balance sheet, not the customer's. Get ready. Customers are going to jump on this model like white on rice.

Shift #2: Complexity's long and illustrious reign will end. Simplicity will be king.

"Masters of complexity"—it's what most tech companies pride themselves on being. But as we pointed out, customers are rapidly losing their awe of complexity. Simplicity is becoming the new hallmark of sophistication. You can thank Steve Jobs for that. This does not mean that complexity goes away; it means that the new "cool" won't be showing the customer all the inherent complexity, much less turning responsibility for it over to them. It will be showing the customer that your mastery of it has resulted in something truly impressive: simplicity.

Like other aspects of the shift, this was waiting in the wings for years and is an expected dimension of the industry's maturation. Electrical power was also a marvel of complexity in its early days. In fact, it still is. But few electricity customers care about, much less want to be a part of, that complexity. They just want it to be there and to work. At least to some degree, this is our destiny too . . .

For their part, customers are becoming resentful of the costs of managing the technical complexity that's handed to them by their suppliers. Complexity is exhausting IT budget dollars that could capture more value if they were redeployed to better ends like new capability, more capacity, or faster adoption. This same complexity reduces the productivity and efficiency of the whole IT department and its systems. And while no one knows the exact figure, billions of dollars in failed IT projects can also be laid at complexity's doorstep.

So while this is hardly new news, the combination of the tough economy and the promise of the cloud have accelerated this trend markedly. One major benefit embraced by cloud providers is marked reductions in entire layers of complexity. It is a well-timed offer, given the economic pressures on the CIO to reduce IT management costs and increase business results.

The cost reduction aspect has many dimensions. First, huge capital outlays are entirely avoided in favor of variable costs over time. Out are step-function increases in server, storage, and network capacity that must be purchased in advance of the actual business requirement. In are easily varied capacity models that rise and fall in near-real time based on a customer's small increases or decreases in demand. Simplicity has shifted us from buying all the IT piece-parts in advance to buying only the capacity we really need at any given time. Once again, while the degree of this varies depending on the layer in the stack that you serve and on the "industrial strength" nature of your positioning, the trend is undeniable.

Second, the overall cost of IT infrastructure is coming down. IT infrastructure is layered with a stack of technology components—everything from servers to databases to routers and application software. Standards at each layer of the IT stack have commoditized what used to be differentiated. As long as a given component is standards-compliant and delivers the promised service level, it does not much matter which vendor it comes from. Standard solutions like TCP/IP for networking or Linux for computing are a great simplifying force, as they remove custom support requirements and enable more vendors to compete to deliver a given part of the infrastructure. This simplification is reducing the costs of many aspects of computing each and every year.

Finally, as we all know, enterprise technology has been criticized as "some assembly required." That is, even if I believe the promise of what the technology can do, there is a mountain of customization, configuration, implementation, process redesign, and training complexity between the purchase of technology and actually getting any value out of it. Imagine giving your son a bicycle for his fourth birthday, and then telling him that it is going to take until he is six years old before it will be assembled and working. The cloud promises to cut this complexity down to size. The entire implementation cycle for an enterprise application could take weeks instead of years. The infrastructure is already in place. Security, redundancy, and backup are built in. Customizations are often done by the end user once the system is live, as opposed to having a central team attempt to guess how the users will want certain things to work.

Another gigantic benefit is that the refresh and upgrading of the technology becomes a nonevent. Just as the apps on your iPhone dynamically update themselves, your enterprise apps will be upgraded constantly as the vendor evolves their solution. Gone are the big-bang, high-risk upgrades (think SAP R2 to R3, for those of you who are old enough). Here are new features making themselves available as options for the end user on a weekly or monthly basis, sometimes even faster.

Shift #3: Cloud customer aggregators will shrink the direct market for tech infrastructure providers.

In the good old days, every company was a prospect for every tech product provider. Every company had its own IT department, its own IT stack, and its own labyrinth of suppliers. We were used to selling big deals to big companies that needed every component of every layer in a complex stack. As tech companies, we had not just 500 buyers in the Fortune 500, but probably 15,000. Divisions and subsidiaries acted independently. We had tons of customers to call on, tons of deals to negotiate and get done. And that was just in the biggest global customers.

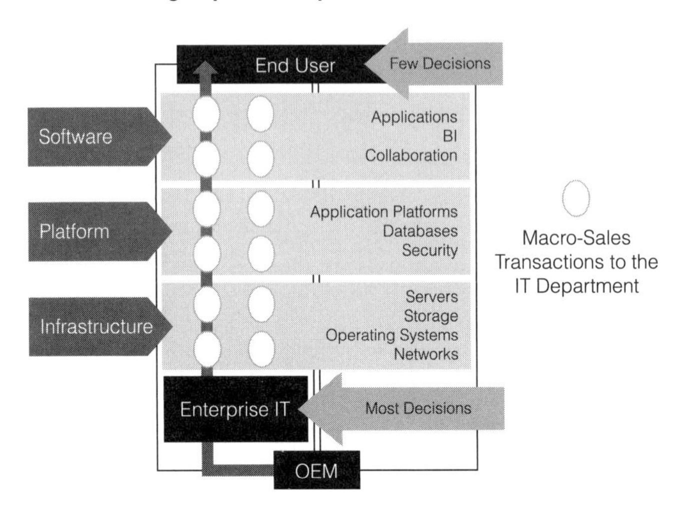

FIGURE 2.5 Selling CapEx to Corporate Customers

Add to these the small to medium-size businesses (SMBs), and the sales arms race was on. Direct models, indirect models . . . it was all about coverage. There were literally tens of thousands of decisions per year per component category, because the ratio of technology platforms to end-customer organizations was at least 1:1. No two companies ever shared the same stack. Every corporate customer was making decisions about what products to use at every layer of every stack they owned. Sales organizations had to be big because the number of deals was big. Occasionally, sales could find deals where there was very little competition. Exciting margins could be had since no one else was around to undercut prices or point out the weaknesses in your offer. Other times, all your competitors were present, and making money was tough due to high sales costs and discount pressures. But the bottom line was that you wanted to be in every deal, and that meant a huge commitment to marketing and sales resources.

How will the cloud change this? Significantly—especially if you are a component provider at the infrastructure layer of the technology stack. Why? First and foremost, it is because the existence of huge, cloud-based infrastructure as a service (IaaS) providers are going to disintermediate a lot of core component providers from selling directly to end customers. If IT departments can buy secure storage over the Internet from industrial-strength, massive-scale, secure IaaS companies like IBM, Rackspace or Amazon, they won't need to mess with 3Par, EMC, and Symantec.

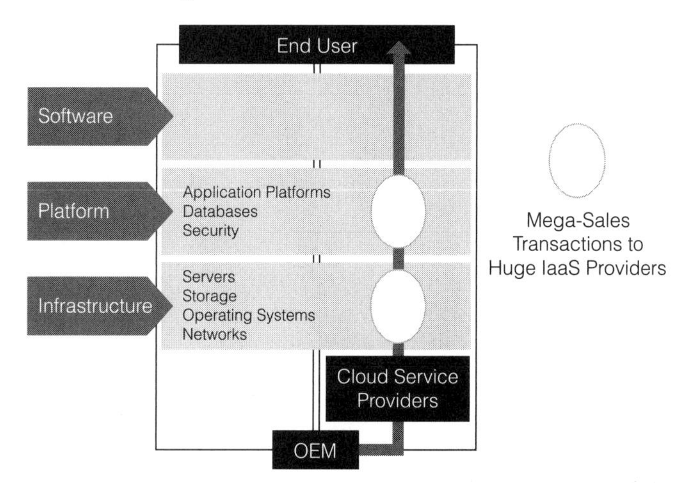

FIGURE 2.6 Selling to Cloud Service Providers

Why would customers deal with all that added complexity if they can simply contract with one cloud IaaS supplier? So if 3Par, EMC, and Symantec don't want to lose the revenue associated with that individual end customer's IT consumption, they have to win the business of the IaaS provider that the end customer chose. At least then they will still get a taste of revenue from that end customer's IT budget. If the IaaS provider goes with someone else, they are out.

There is some good news here. The big IaaS guys will likely crowd out the little IaaS guys who cannot achieve the same economies of scale. While there will certainly be companies attempting niche IaaS plays, like IaaS optimized for electronic medical records, you have to believe that scale will win the day in this business. That means there will likely be a small number of huge customer aggregators. Since there aren't that many players, EMC won't need as many salespeople to call on them. The other positive is that those deals will be huge! These mega-transactions will make or break some component providers' entire fiscal year. That's how important winning the IaaS deals will be if you sell infrastructure technology.

But there is bad news as well. These deals will be unbelievably competitive. The buyers (the IBM or Amazon IaaS providers) will be very sophisticated. They will have every possible component provider on the bid list. If your company is one of them, they will demand that your component solution be massively industrial-strength. They will demand R&D customization and special services, and then they will negotiate the price relentlessly. The gross margins in these deals will be razor-thin.

Even if you win one of these deals, these aggregators will have an interest in minimizing your differentiation over time. They will bury your brand in their solution unless you prove it is in their best interest to do otherwise. And if you can't prove that your differentiation will increase their market share, they will push for solutions based on industry standards, making market adoption of your "standards-plus" features tougher and tougher over time. These market-making aggregators also will be working to dual-source your component or even design it out. The end result will be ongoing pricing pressure for both your products and services.

One other thing: These "customers" also will be your competitors. They will be out chasing all the IT department deals that your direct sales force has in its sales forecast. They will try to break the web of favors that your salespeople have built up over time with their customer counterparts. They will be looking to lock in proprietary advantages for their offers. This could come from three sources. First, their M&A teams will be on the hunt to acquire all the same great new products and companies in the same categories as your M&A guys are. Second, they will be pressing their component providers for exclusive access to differentiating capability—maybe even to the exclusion of the component provider's own direct customers! And finally, they will be taking all your best technology and trying to develop their own differentiating R&D that they can include in their offer. They will act monopolistically—and if you are in the infrastructure component business, you will play ball.

In short, these IaaS mega-transactions will be tough business: huge revenue volume, but tough to win and very low margins. Perhaps even more importantly, your direct business will shrink.

Shift #4: Big changes will come to the channel ecosystem.

As we all know, most tech companies rely on a channel of product resellers to serve some key segments of their markets. Two categories of partners—at opposite extremes of the market—often dominate the landscape: the global system integrators (GSIs) and the SMB-facing value-added resellers (VARs) and dealers.

The GSIs are important because they are often the primary owners of certain large deals. The classic case has always been large government purchases where a prime contractor (an SI) was awarded a long-term contract to piece together the components of a technology system by sourcing and integrating subcontracted products and services. Similarly constructed arrangements have become common in large enterprise markets when Fortune 1000 companies don't want to act as their own "general contractor" to build out a solution. The GSIs are particularly well suited to multivendor projects where no single technology company could provide the entire solution on a direct basis. And since they are perceived as product agnostic, GSIs are trusted to select best-of-breed products across all parts of the solution.

GSIs like Accenture, Capgemini, Deloitte, and Infosys are typically very large and well connected in these lucrative customer accounts. If the tech companies don't play nice with the GSIs, they simply don't get sourced to be the component provider for those deals. This often means slicing out a healthy chunk of product margins as incentive and agreeing to let the GSI capture nearly all of the services associated with the deal.

At the opposite end of the market, tech companies rely on SMB-facing VARs/dealers to perform exactly the same functions, but for entirely different reasons.

We embrace the GSIs because they often control large, profitable product deals that we would love to handle on a direct basis, but simply can't win, so better to have some role than none. But we embrace VARs/dealers because they sell small deals that could never be profitable for the tech companies on a direct basis. So while both major categories of resellers perform the same key functions of driving the sales process and providing needed implementation services, the motivations for the tech companies to woo them are totally different.

So let's break this down at one more level of detail.

Global System Integrators vs. VARs/Dealers

Reseller Category	Customer's Product Business	Customer's Service Business	Reseller's Primary Incentive	Reseller's Service Skills	Reseller's Business Expertise
GSI	Desirable for tech company to go direct	Desirable for tech company to go direct	Professional service revenue	Sophisticated global multivendor	High
VAR/Dealer	Undesirable for tech company to go direct	Undesirable for tech company to go direct	Product and maintenance revenue	Basic, local	Low

FIGURE 2.7 Global System Integrators vs. VARs/Dealers

While their motivations are different, the tech company is awarding strong margin incentives to its resellers in both categories. The VAR/dealer channel is about a high volume of little deals, of which the tech company only receives a small gross profit in absolute dollars. But get enough VARs/dealers to do enough small deals, and you end up with, well, Microsoft. Think about the huge franchises that have been built on the back of this model: HP's printing business, Canon copiers, and Sage software, just to name a few. So with the margin on products going down and the cost of sales going up in sector after sector, why not give as many complications as you can to your channel partners and just be happy with your small product gross profit?

Where are the shifts in this well-worn model going to take place and why? We would argue that:

- Tech companies will begin to market their cloud offers directly to customers large and small. There will be massive channel conflict.
- Cloud offers will require fewer of the basic technology services that are the bread and butter of many channel partners.
- Vertical market focus and business process expertise (via consulting, managed services, and outsourcing) will become the new high-value service capabilities. The Consumption Gap is already fueling that trend, cloud or no cloud. Most VARs/dealers just don't have those skills today, but the GSIs might. This will make the GSIs an even more powerful force. At the other end, the lack of these critical skills could challenge the small VAR/dealer's ability to play a role in the value chain at all.
- IT departments will be worried about their end-user usage data flowing through a channel to multiple players in an ecosystem. But whoever has access to that data and translates it into real business value will become the trusted advisor to the account. That could mean even more direct customer emphasis and more channel conflict.

This combination of these shifts will impact both kinds of resellers, but *much* more so to the SMB-facing VARs/dealers. If the cloud enables the tech company to achieve SMB revenue directly—without the "middle man"—and the basic technical services required to install and configure the solution are not required, why will the tech company want to give up the margins? They won't.

Reaching the SMB customer used to be a mess for large tech companies. Now the cloud is making that whole segment of the market a lot simpler to reach. This will drive many more tech companies to be aggressively direct in that segment and will disintermediate a certain percentage—in the case of software, a large percentage—of resellers.

The best will survive, but retooling to excel in the new model is no simple task. For VARs/dealers to survive, many are going to have to get into new, more complex offers like managed services or business process consulting. Their niche is still their ability to put local labor into a small account. That used to be to sell and install products. But now the SMB customers can do those themselves. They can get just about all the IT capability they need virtually in the cloud. They already have PCs, a smartphone, and a printer. The additional computing power, storage, software, and data services might be all in the cloud. The one thing they can't get in the cloud is the attention of a consultant who can diagnose and solve a business problem. As we pointed out, this is often *not* the native skill set of small tech resellers.

The GSIs will be impacted by direct cloud product offers in a way that creates both threat and opportunity. The GSI's customers might buy more of the technology directly off the cloud and disintermediate the GSI out of its traditional cut of those product sales. Big customers and government accounts are also expecting far less technical complexity and better data compatibility from the cloud, and thus expect to spend far less on the services typically associated with it. So while the cloud does threaten some of the GSIs' usual revenue and margin sources, they will just need to plan for a mix shift away from the technical projects and toward more consultative ones. This is not fatal, as most of the GSIs already have a combination of technical and business skills in their workforces. If they succeed at this transition, they could become even more influential because they will still have great relationships with certain large, profitable enterprise and government accounts, they will have the higher-value skill sets that are in demand, *and* they are product and vendor agnostic.

But the big and interesting possibility is that the GSIs might be large enough and trusted enough to get permission from the IT department to access their usage data from public and private clouds—maybe even from the IT department's internal servers. They may succeed in getting their large customers to force the tech companies that provide the clouds to give the GSI their data as well as the analytics needed to drive the adoption of end users.

As we talk about throughout the book, whoever performs this function is likely to become the customer's trusted advisor and to be in the pole position to win future business. While the VARs/dealers often lack the brand and capability to successfully assume that role for the customer, the GSIs could pull it off. Imagine what a coveted space they could occupy if they could optimize adoption across whole systems or departments by combining the data from multiple product vendors or across various clouds to get an enterprise-wide view!

What's critical is that large parts of the reseller ecosystem need to begin to transition their "value add" from technical expertise to business expertise. This means:

- Vertical industry expertise (health care, legal, financial, etc.).
- Functional expertise (accounting, HR, marketing, etc.).
- Solution design expertise.
- Analytics expertise.
- Consumption expertise.
- Business process improvement and change management expertise.

They will also extract their margins less from products and product-enablement services and more from services like consulting, managed services, and outsourcing. These are new and different business models for many resellers but these transitions are critical to the ongoing viability of their companies.

Shift #5: Cheaper enterprise software will emerge.

We all know about 99-cent software in the Apple App Store. But what does this mean to enterprise software?

In December of 2008, we held a TSIA Executive Board Meeting in Palo Alto, California. At that time, the world was a little more than a year into the iPhone craze. A really smart board member made the off-hand comment that he didn't think that the existence of good, free consumer software was going to do much to help the enterprise software market.

Fast-forward to the spring of 2011. We were having a roundtable discussion of 30 senior execs from the largest tech companies in the world. We asked them this question: "Who believes that enterprise software prices will go up in the future?" Guess how many hands went up? Zero. Nada. When we flipped it around, there was near unanimous agreement that prices were going to come down.

But wait, you say. There is a world of difference between industrial-strength, mission-critical, secure software like SAP ERP 6.0 and consumer apps like Facebook. No doubt that is true. But do you remember who Ken Olsen was? He was the CEO of Digital Equipment Corporation. DEC and its VAX midrange products were an 800-pound gorilla in the 1980s world of computing. Along with IBM, they dominated the market for large-scale, mission-critical enterprise computers and operating systems. Then an upstart called Sun came along with low-priced offers based on Unix and RISC. Mr. Olsen never saw Unix as a challenger to his "serious" VAX/VMS systems. "Snake oil," he famously called it. And it turned out he was partly right.

Unix did not end up being the true victor, the true replacement for big, bad VMS. It turned out to be Intel + Windows and then, eventually, Intel + Linux. This was not technology that was one rung down the "serious computing" ladder. It was the bottom rung, the very opposite end of the ladder. Over a period of just 20 years, and at a time when the pace of innovation was glacially slow compared to today, industrial-strength computing became the domain of companies that grew up selling to consumers. Mr. Olsen was focusing on the wrong threat.

Well, here we go again. The new world of software looks to be smaller, more focused, highly modular, and cheaper. With history as our guide, we believe that industrial-strength attributes will appear in low-cost software over time. Data will begin to move seamlessly across this next generation of applications. They will become more secure. There will be cheap apps for business. There already are.

To see how rapidly things can change, let's go back to our CRM example of Siebel and Oracle. In the early 2000s, Siebel was dominating Oracle in the CRM market, with annual revenues of around $2 billion. After years of frustration, Oracle just bought them. Today Oracle charges around $300 per month per user for Siebel CRM Base (according to a published price list), and then supplements that fee with dozens of functionality-based upgrade options. As you will recall, salesforce.com's top published price was $250 per user per month, just under the Oracle Siebel CRM Base price, for "Unlimited Functionality." Think that's the end of the price war? Think that Oracle will be able to make the "industrial strength" argument to prevent continued commoditization? We don't think so . . .

In June of 2011, salesforce.com announced that a huge U.S. government agency had decided that its technology was more than a toy.

Salesforce.Com Achieves GSA Moderate Level Authority

By Madhubanti Rudra, TMCnet Contributor

The leading provider of on-demand, Cloud based CRM solutions, Salesforce.com recently announced that its services have received U.S. General Services Administration moderate level Authority to Operate. The certification is based on testing performed against the NIST 800-53 Rev. 3 moderate baseline requirements consistent with FISMA requirements, the company clarified in a press release. With this, salesforce.com hopes to bring the same customer success to the public sector that it has been delivering to the private sector for the last 12 years.

Salesforce.com has received the Authority to Operate at the moderate level baseline for the Force.com platform, Salesforce CRM, Salesforce Chatter, and supporting infrastructure. The certification implies that government customers will now be able to leverage salesforce.com's cloud computing services to address even more of their IT needs, all without any hardware or software.

Salesforce.com is moving up the seriousness stack. But wait! They are not the low-priced CRM company anymore! While salesforce.com offers a range of functionality levels, the $65-per-month Professional level is a popular one. Well, here comes Microsoft Dynamics and its shot across the bow at that same level of user! Let's try $44 per user for CRM.

Microsoft Dynamics CRM Pricing

	Salesforce.com			Oracle	Microsoft
Features See details below	**Professional Edition**	**Enterprise Edition**	**Unlimited Edition**	**CRM On-Demand**	**Dynamics CRM Online**
Office Fluent User Interface	✘	✘	✘	✘	✔
99.9% Service Level Agreement	✘	✘	✘	✘	✔
SharePoint Sites Interoperability	✘	✘	✘	✘	✔
Complete Offline Solution	✘	✘	✘	✘	✔
Presence Management Interoperability	✘	✘	✘	✘	✔
Order and Invoice Tracking	✘	✘	✘	✘	✔
Monthly Cost	**$65 USD**	**$125 USD**	**$250 USD**	**$75 USD**	**$44 USD**

✔ Included ✘ Not Included

Source: Microsoft website, July 2011. © Microsoft.

FIGURE 2.8 Microsoft Dynamics CRM Pricing

"Promotional pricing of $34 USD per user per month is available through June 30, 2011."

See below for important information about the results.

The Microsoft Dynamics CRM Online Cost Comparison Calculator provides a customized estimate of potential cost savings.

- **How Microsoft is able to provide cost comparison rates:** The cost comparison report is fed from publicly available information/pricing posted on each competitor's corresponding websites. This calculator information/pricing is checked regularly. The results do not constitute an offer. Contact us and we will gladly help you calculate the potential savings.
- **Things to remember about the comparison:**
  - Eligible promotion pricing and discounts are not reflected in this calculator.
  - Microsoft Dynamics CRM Online pricing is $44 (USD) per user per month. Promotional pricing of $34 (USD) per user per month is available through June 30, 2011. **Please note that this promotional pricing is not reflected in the calculator results.**
  - Contact Manager, Group, and Force.com editions are not included in this comparison as there are no comparable editions in Microsoft Dynamics CRM Online.

Source: Microsoft website, July 2011. © Microsoft.

FIGURE 2.9 Microsoft Dynamics CRM Promotional Pricing

And darn! We just missed the summer sale price of $34!

Microsoft is undercutting both salesforce.com and Oracle in the CRM market. And just to prove the cloud is a fair world where no one is immune to commoditization, look at what is happening to Microsoft's traditional cash-cow office apps market!

GSA Entrusts Email Service to Google's Cloud

By John K. Higgins, E-Commerce Times

In a move that confirms the U.S. government's recent emphasis on seeking Cloud-based information technology solutions, the United States General Services Administration has awarded a contract for converting the agency's email system to a Cloud platform. GSA gave the US$6.7 million, five-year task order to Unisys (NYSE: UIS), which partnered with Google (Nasdaq: GOOG), Tempus Nova, and Acumen Solutions to compete for it.

"Cloud computing has a demonstrated track record of cost savings and efficiencies," said Casey Coleman, GSA chief information officer. "With this award, GSA employees will have a modern, robust email and collaboration platform that better supports our mission and our mobile work force, and costs half as much," she said.

The GSA is not just looking at saving a few bucks on the licenses, they are buying the whole cloud-based story hook, line, and sinker. Not only do they see the cloud as cheaper; they see it as better.

The article goes on:

While some federal agencies have moved parts of their organization's email systems to Cloud platforms, the GSA contract marks the first time a federal agency will adopt a Cloud-based system for email on an enterprise, or agency-wide, basis. The migration will result in a 50 percent savings over the next five years when compared to current staff, infrastructure, and contract support costs, GSA said. Nearly 17,000 GSA users will be affected by the change.

GSA's decision to move to the Cloud occurred in the midst of a policy change spurred by the Office of Management and Budget to seek more dynamic and efficient IT solutions. In a report on its email capabilities, GSA said that its existing infrastructures were "not adequate" for the future.

"They consist of aging, site specific servers, with limited redundancy and inconsistent archiving capability. In addition the current email system makes it difficult to manage and retrieve emails associated with legal matters. In-house upgrading and replacement will be both expensive and disruptive," GSA says in a bid invitation briefing document.

In conformity with a May 2009 White House directive, GSA explored innovative alternatives and found that "traditional outsourcing and system integration support is insufficiently adaptive and costly and should be replaced by commodity services with a Software as a Service (SaaS) Cloud computing strategy," the briefing document says. GSA's current system relies largely on an IBM (NYSE: IBM) Lotus Notes and Domino software platform.

The agency's needs caught the attention of Vivek Kundra, U.S. chief information officer. "GSA's current environment lacks the level of integrated features that are commercially available. GSA requires a greater use of features such as integrated messaging and collaborative tools to support its mission. The storage associated with e-mail archiving continues to grow and is costly to manage," Kundra said in a blog post.

GSA's requirements appear to dovetail with the features available through Cloud computing. One of those features is scalability—the capability to add users in a cost-effective manner. GSA's system, for example, may ultimately be expanded to 30,000 users, according to its contracting documents.

The social networking capability facilitated in a Cloud system was also attractive. GSA required that an updated email system provide an effective collaborative working environment. In terms of cost controls, GSA sought to reduce in-house system maintenance with system tools that provide business, technical and management functions. Lastly, the agency required appropriate security and privacy capabilities.

"This decision is a clear signal that the federal government is headed to the Cloud. Google Apps has seen incredible momentum at the state and local level—and we're excited to see that same kind of rapid adoption on the federal level as well," David Mihalchik, a business development executive at Google, told the E Commerce Times. "Other agencies will be looking closely at what GSA has done."

The system will feature several applications including Google Gmail, Google Calendar, Google Docs and Google Sites. GSA will tie all of its 17 locations, including its international offices, into the system. The email migration will begin next year.

"Earlier this year, Google Apps became the first suite of Cloud computing email and collaboration applications to receive Federal Information Security Management Act (FISMA) certification, enabling agencies to compare the security features

of Google Apps to that of existing systems," Bradshaw added.

Just six months after this article appeared, the momentum for low-priced, effective cloud office tools within the U.S. government took another huge leap forward. If you thought the 17,000-user deal above was exciting, how about the 950,000-user deal that is up for grabs as we write this book?

April 28, 2011

U.S. CIO Vivek Kundra said federal agencies will use the Cloud to consolidate 950,000 email boxes now sitting on 100 different systems.

By Elizabeth Montalbano, InformationWeek

The General Services Administration (GSA) will unveil a $2.5 billion procurement May 10 to consolidate email on Cloud-computing infrastructure as part of the federal government's "Cloud first" mandate, President Obama's CIO said this week.

Federal agencies have identified 950,000 email boxes across 100 systems that can be consolidated by leveraging Cloud computing, and the procurement will be to implement this plan, U.S. CIO Vivek Kundra said Wednesday, speaking at a forum of federal IT officials.

Didn't think that salesforce.com and Google were moving up the seriousness stack? They are. Just like Intel did to DEC. Google just took a 30,000-seat deal away from IBM. The cloud is bringing software prices down. It will be easy to convince yourself that it is not going to happen to your company or that it is not going to happen on your watch, but there are growing signals that this phenomenon will be real and pervasive. Who is your disruptive competitor tomorrow?

If this shift continues and software prices do decline precipitously, there is a risk that the cloud could drain the profit pool in a lot of software markets. That is a potential downside for sure. But on the flip side it could also dramatically increase the size of your market.

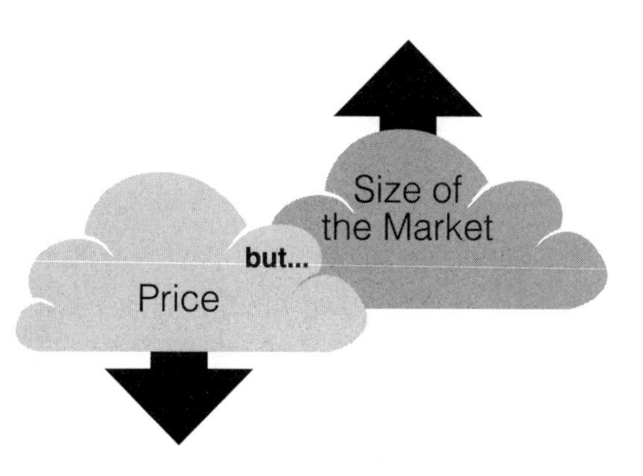

FIGURE 2.10 More or Fewer Total Revenue Dollars in the Cloud?

Lower prices are also certain to move more people into the market. Apple is proving that low prices are powerful drivers of volume. They are convincing app developers that, while cutting software prices by 90 percent may not sound like much fun, it also might increase their unit volume not by 10 times, it might increase them by 10,000 times.

Equally likely to increase the size of the market for cloud software is the ease of ownership. By eliminating complex installs, upgrades, and numerous other PITAs, software companies can bring more "main street" customers into the market for their products. This is a huge opportunity for tech product companies that have heretofore been limited to selling to technically savvy, early adopters. For every one of those folks, there are 100 of the pragmatists waiting for lower prices and a better experience. This is not just a consumer story, it is a B2B story too. Low prices but huge volumes may not mean less water in the profit pool, it just means that the rules of the pool are changing.

Shift #6: IT departments will "get out of the way" of end users.

Remember our diagram showing where most tech companies focus their sales efforts?

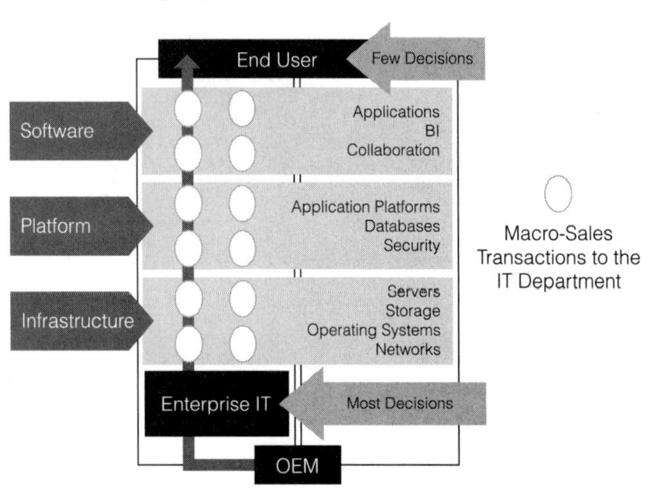

FIGURE 2.11 Selling CapEx to Corporate Customers

On virtually every technology purchase decision of any size, the IT department historically played a controlling role. If the product and supplier decisions were related to the core infrastructure of the company's systems, data, networks, security, or performance, IT owned the whole decision. If the technology involved was new in any way, IT made the call. Even if the deal was for just more of the same technology they already used, IT gave the nod to the procurement department before negotiations with the supplier commenced. This control frequently extended all the way out to the end user. Many companies dictated which PCs their employees had, which phones they bought, and certainly which software they used.

Here again, the rules are changing. And once again you can thank Steve Jobs, Mark Zuckerberg, et al. IT's dominance is giving way to the growing influence of the end business user. While it is not a new trend that the business users are growing in influence, up until now, the "business user" reference really meant top functional execs rather than actual end users. The CFO's senior team weighed in on F&A apps. The sales vice president's team weighed in on CRM apps. And when it came to ERP and other manufacturing-related technology, the development and operations execs led the way—much like senior doctors do in choosing medical equipment. But the key distinction is this: The technology decisions were centralized at the top, and IT played a key role. Even when IT could not say yes, they could often say no. The actual end user was nowhere to be seen.

Because of these dynamics—large corporate tech deals that were decided by high-ranking centralized buyers—tech companies rightly played the game according to the rules set out by the customer. RFPs were drawn up, suppliers vetted, decision makers courted, deals negotiated, contracts signed, systems implemented, and usage mandated—all by centralized authorities. The last person to find out? The end user.

As we have already mentioned, this process has led to the Consumption Gap. It was an old-school model that was the best we knew how to do within the confines of the customer's procurement process. The Consumption Gap that was caused by centralized technology decision making might have been right when tech was complex, risky, and expensive, but it is looking more obsolete by the month.

Today, the power is moving from the center to the edge. End users, emboldened by their success with their personal technologies and loyal to the models and suppliers that provided them, have begun taking their business computing fates into their own hands. They are not just picking their preferred phones and PCs; they are often choosing nonstandard, nonsanctioned software for use in the course of their jobs. They Tweet, Google, YouTube, GPS, Excel, and LinkedIn their way to better business performance. They pick what fits and what's fun . . . and they actually *use* it.

Their choices are now invading the workplace. In rare cases, users are even banding together to overthrow centralized IT decisions. And it is working. The traditional enterprise customer's dominance by the CIO and the IT department is slowly giving way to the growing influence of the end business users. Name a consumer technology company that does not have one eye on the enterprise market. They see the trend. Apple, Amazon, Google, etc. are circling. At some point, the central authority begins to give in. More and more wise IT leaders are not just succumbing to the trend, they are cheerleaders for it.

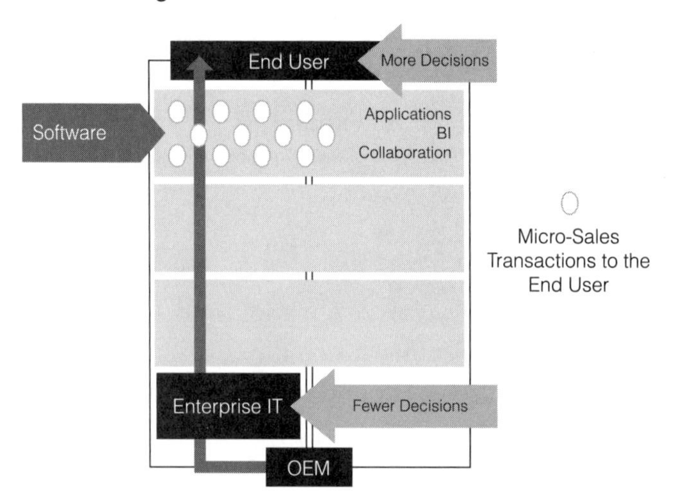

FIGURE 2.12 Selling Software to End Business Users

But let's talk about how foreign the relationship between some tech companies and their end customers can be. We could build a case that the percentage of total interactions between major enterprise technology companies and the end users of their customers could be close to zero today. This includes all the interactions that make up the end-to-end life cycle between the tech company and their corporate customer. Do they touch the end users in the sales process? No. What about in contracting and collections? No.

What about in installation and implementation? No. Well, surely they do that in education and training? In fact, it's becoming less and less often today. "Train the Sys Admin" is usually the cost-saving maneuver that corporate IT buyers settle on, if they pay for training at all. Okay, you say, but they *must* touch the end users in the customer service and support process! Well, again, it's not often. It is usually the customer's internal help desk that deals with the end user's problems and usage questions. It is the help desk, in turn, who actually interacts with the tech company.

So by design, there are many large enterprise tech companies that have very little experience in dealing directly with end users. In the summer of 2010, we attended a large conference of technology resellers. On the stage were executives from five large enterprise tech companies. They were talking about the importance that the resellers played in extending their companies' product distribution out to the small and mid-size companies that they could not profitably reach through their direct sales force. They also talked about the critical need for resellers to provide the services needed to make those SMB customers successful. At one point, one of the executives starting talking about how important product adoption was, but he cautioned about the limits of that from a practical standpoint. "No one wants to touch the end user!" he said. And everyone on the panel laughed.

Now again, it is not because these executives or their companies are insensitive to end-user success. Far from it! They know the importance of end-user adoption, but their companies simply have no facility, capacity, or business model for dealing with end users in any part of their customer relationship. The economics of the traditional go-to-market models for enterprise tech companies simply never contemplated a serious role for the end user in revenue generation. But now, we are facing the rise of the end users. They will be making the hundreds of thousands of micro-transaction purchase decisions—the little dots—that will drive the billions of dollars in corporate IT spending that an enterprise tech company covets.

Things have got to change. Enterprise tech companies are losing ground every day to the Apples, Amazons, Googles, and salesforce.coms who built their entire business models around serving end users. This is one of the main points of this book. Enterprise tech companies need to learn to love the end user, and be able to do it on a massive scale.

If giving end users the flexibility they need to find technology solutions that make them more productive, then vive la revolution! Of course, there needs to be control, and clearly not every tech decision can be left to the whims of individual employees, but the trend is evident, likely to accelerate, and has strong historical precedent in the brief history of the computer industry. Soon we will set end users free. Individual users of cloud apps like salesforce.com are already voting with their keyboards and touchscreens. They are selecting the aspects of the platform that are most important in helping them to do their jobs. The next stage will be enterprise cloud apps that provide highly configurable modules and interfaces for end users. As with the iPhone, users will select, combine, and tailor specific modules to meet their unique needs and work styles.

Shift #7: Tech companies will capitalize on user-level behavioral data.

This shift, perhaps the least talked about of them all, could prove to be the most exciting and game-changing. Why? Because it could potentially eliminate the Consumption Gap and inject whole new streams of value into tech markets old and new. For the first time, it may become possible for tech companies to successfully drive end-user consumption of product value through real-time interaction. This could become a key part of the cloud's legacy and could power a new front on the war against commoditization: we call it the Consumption Model.

The true power of the cloud may turn out to be that the end user and the tech company are connected in real time, and the usage activity data no longer resides on the corporate customer's servers, but now resides on your company's servers. Every click of every mouse on every screen in every module can now be known. The things that can be done with that data could dramatically enhance your company's value proposition. Soon we will tailor the presentation of the technology product to the specific needs and preferences of individual users in real time.

Soon cloud companies will aggregate usage data to discover best-practice utilization patterns. We will build insights and recommendations that will improve the user's personal or business outcomes. We will find ways, directly and with partners, to insert this new and relevant content into the user experience at just the right time to improve the product's use. And finally, we will give business and IT management unprecedented insight and influence into how their employees are adopting, benefiting from, and even innovating through the technology tools they have been provided.

We have asked about the existence of this "raw material" data at dozens of TSIA member companies that are all or partially cloud-based providers. The answer is always some form of yes. But the irony of it is that this data—what we will argue could be the most game-changing single asset of the next-generation enterprise and consumer technology company—is usually sitting idle.

The potential impact of this data and the real-time end-user connection on the sales, marketing, services, and development investments of enterprise tech companies will separate the winners from the losers. We are headed toward a future of end user-driven, micro-transaction account growth models. We will cover this in much more detail later in the book.

Add Them All Up, and . . .

These seven shifts could profoundly impact existing tech company models. Not everyone will like them; not everyone will want them. In fact, many existing tech companies will fight against them tooth and nail. Some will attempt delaying tactics to defend the old rules that shelter their current revenue and profit streams until the day they can cash out and retire. They will highlight the risks that customers could incur, which might lead to disruptions of their business operations or potential security lapses if they switch to a cloud competitor. They will exploit their advantages as a proven, mission-critical provider. They will also point at the customer's sunk cost. The customers may have invested millions of hard dollars, and many times that in internal costs, to get their current IT systems working as they are today. The incumbent tech companies will lean on the risks of giving all that up for some upstart computing model.

Just like Ken Olsen and DEC did.

So where is this all headed? No one knows for sure. New economics are coming to the technology industry. They are coming fast, and with them will come new rules. How tech companies succeed and why they fail may never look the same again. The tools to deliver value have never been this exciting before! It could truly be time for a "new normal."

The New Normal

What's in Trouble	What's in Vogue
• Complex technology	• Simple
• Asset purchases	• Pay for results
• Install, integrate, upgrade	• Business value
• Maintenance costs	• Turn on, turn off vendors
• Hard-to-use software	• Higher volumes
• Central IT control	• End-user preference

FIGURE 2.13 The New Normal

We will look at some of these shifts and the new normal in coming chapters of this book. But first let's take a deep and important look at the way commoditization plays out in the unique world of technology products. We are not selling wheat futures or pork bellies here. Commoditization in the "old normal" of tech is not a simple story. We have a perspective on the Consumption Gap's role in causing it and customer lock-in's role in preventing it. This perspective might not just open your eyes to the true effect of category commoditization in your markets, it may also give you unique insight in how to reverse the trend.

3 Looking Over the Margin Wall

Tech is not immune to commoditization—it's just harder to spot. The phenomenon occurs when one product becomes close enough to its competition that customers start to buy on price alone. Since tech products are usually complex and feature rich, it is hard for the company that makes them to imagine that they would ever reach such a lowly state. Yet more and more product categories—hardware, software, even advanced products like medical devices—seem to be feeling the pinch of price competition and customers who seem increasingly indifferent to our latest attempts at differentiation.

There is a point where the company just can't afford to continue with the product. The prices are getting too low. Yes, even tech products—entire categories—can hit the Margin Wall.

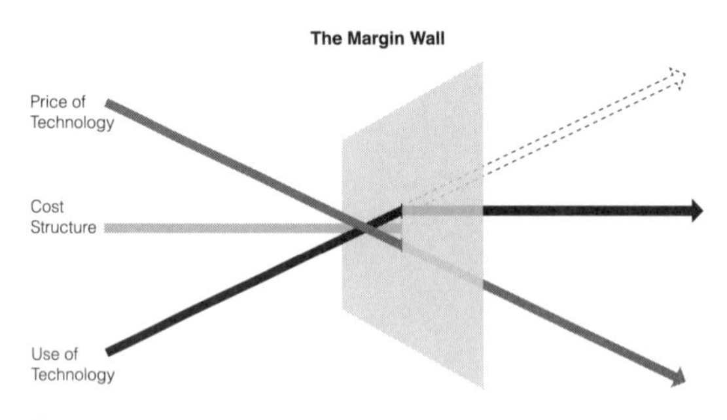

FIGURE 3.1 The Margin Wall

You'll know your product is at the Margin Wall when your cost of goods on the equipment plus your sales, marketing, and service costs on the deal add up to more than the customer paid for the system. Once you get there, it doesn't seem to matter what you do to the product, how many new things it can do, customers just won't pay a premium. It is frustrating to the tech company and its engineers because they know that the product can do some very cool things better than the competition. Maybe it can do something that the competitor's product can't do at all! But the Consumption Gap, the evil enemy of differentiation and margin, has set in. Despite all your exciting features, the customer is still just going to use the product to do the basics. They know how to do them, but that's all they know how to do. So when they go to buy the next product, they choose the one that can do those basics at the lowest price. When that happens, the whole product category crashes into the Margin Wall. Commoditization in tech is real—maybe more real than we have admitted.

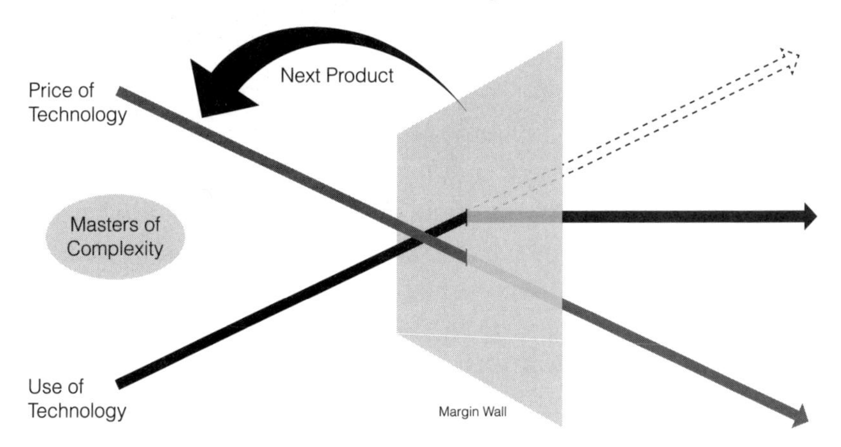

FIGURE 3.2 The Product Playbook

But no worries! We know what to do! We have been doing it successfully for over 40 years. We run a play from the product playbook. The concepts in the product playbook are well understood by any good tech company exec. We know what to do and what steps to go through to create a successful product: Either tackle a complicated problem, create an innovative capability, or achieve a price-performance target, hopefully one that has never been achieved before. Build a product with all the necessary features to accomplish the breakthrough—maybe even a little overshoot just to be sure no customer objections come up. Then charge the highest price you can. As the early adopters start to use your product and the buzz begins to grow, service the increased demand by expanding your distribution and keeping your prices close to those of your emerging competitors. Once your product crosses the chasm, start focusing on building a profitable maintenance business that will tide you over until you can build the next-generation offer. So when one of our products does finally hit the Margin Wall, we know what to do! We run the same play again.

Fortunately for us, the complex nature of our offers has meant that the runway to commoditization was often a long one—hopefully, longer than our remaining career timelines. Some product categories that ran this play, especially before the rate of innovation was as fast as it is today, took a decade or more to hit the Margin Wall. The Margin Wall is the turning point for tech companies. As an industry, most tech companies are simply not optimized for high-volume, low-margin product businesses. That is what grocery stores do. We are the masters of complexity. That is what we do. We innovate, charge a lot for that innovation, and when and if that innovation ever commoditizes and hits the Margin Wall, we do the smart thing, the only thing our cost structure allows us to do: We come out with another complex, high-margin product.

Fortunately, we also do one other thing that mutes the impact of crashing into the Margin Wall. Even when products become commodities, we keep customers "in the family" by leveraging their huge sunk cost in the last product to keep them locked into maintenance (or supplies or parts or whatever the high-margin aftermarket business is). It works. Nobody is going to heaven because of it, but it does work. Then we come out with the "next-gen" product in the same category. We start right back on tackling a new price-performance target or adding whizzy new features in hopes of creating yet another *Wow!*-compelling offer. Finally, we try to get the customers we kept "in the family" to upgrade to this next-gen offer. We make the upgrade less painful than switching to someone else's product. This practice has a name, it is called "technology refresh." It is a cycle that happens for most tech product companies every three or four years.

Unlike most any other product industry, if a tech product, a whole category, or even an entire technology wave commoditizes, high customer switching costs often save us from disaster. Car manufacturers can't be assured of brand loyalty due to that. But lots of us can. We would argue that this, not competitive feature differentiation, is what has hidden the true effect of commoditization in lots of enterprise tech sectors. The cloud's low switching costs could threaten this critical component of our defense against commoditization.

So when you ask most tech executives what happens on the other side of the wall, for these reasons, not many of us actually know. If you had a stint in HP's printer business or at a cellular carrier, then you might know. One thing that is for sure, almost no enterprise software companies know by experience. Because most of these products were highly complex and because all the software companies had similar business models, few of them used low price as a differentiating strategy. Combine that with the close-to-zero marginal unit costs of software, and the popular belief was that the enterprise software business was permanently profitable and virtually immune to commoditization. But is that really true? If these companies agreed that the new license business would cover its fair share of R&D costs, in addition to the standard marketing and sales costs, and apply it against their new license revenues . . . is it really very profitable to sell enterprise software today?

According to Oracle's latest 10-K filing, they recorded $9.2 billion in new software license revenue in FY 2011. Their sales and marketing costs on that revenue were a little over $5.4 billion. That left a gross margin of $3.8 billion or 41 percent. R&D costs were $4.5 billion. (Note: Those R&D costs were separate from the $1.3 billion in costs for software license updates and product support.) If you just take half of that R&D cost ($2.25 billion) and throw it against the product revenue, you get a new software license business that made a gross profit of $1.5 billion on revenue of $9.2 billion. That is just a 16.3 percent gross margin activity. And this is a company that owns one of the all-time great product franchises. Seventy-two percent of that revenue was around its franchise database and middleware products; just 28 percent was from the sale of new application product licenses.

What would happen if a Google or salesforce.com began to erode Oracle's new license prices by 30 percent? Can't happen, you say? Won't happen, you say? The folks at Siebel/Oracle are probably less skeptical today than when Marc Benioff first started salesforce.com or when NetSuite started to put low-cost ERP into the cloud. These software as a service (SaaS) and Internet companies are dead serious about disrupting the old-guard enterprise players. They are eating around the edges, probably starting on the low end of the markets and working their way up, adding features and mission-critical attributes as they go. Just like Intel did to DEC.

How many other companies, many in the start-up phases, maybe even still unknown to the incumbents, are planning to run a new play from the cloud product playbook? One that is designed from the beginning to capitalize on the vulnerabilities that some of today's leading enterprise software companies have around price, the Consumption Gap, and the limits of change imposed on them by the financial markets. As we asserted in Chapter Two, these new companies might not be just a little lower priced and *a little* cheaper to own and operate—they could be *a lot* less. At the extreme is a company called Spiceworks that offers IT network management software for free! They have raised $16 million in venture financing for their advertising-based business model, and they made it to *PC Magazine*'s Best Product List. It must work well, because they claim that over 1.5 million IT professionals are using it. Even young SaaS companies who are founded on new and separate functionality areas, like Success Factors, will at some point consider moving into more traditional adjacent markets—in this case, HR apps—that are held by incumbents like PeopleSoft/Oracle. Did you know that salesforce.com already has a SaaS product called Database.com?

Commoditization is probably at work in tech today to a much greater degree than we realize. Even if a tech executive does realize it about their own company's products, they are not likely to admit it publicly. The fact is that software maintenance revenues are all that stand between many enterprise software companies and the Margin Wall. What would happen if customers started getting really angry about maintenance prices and had another way out? What would happen to enterprise software companies then? For Oracle, that is a $15 billion, 86 percent gross margin business. SAP has $6 billion in maintenance. Think these are not ripe targets for both SaaS companies and cost-conscious enterprise customers alike?

We believe that hitting the Margin Wall will soon be a real concern for all tech companies, not just hardware companies. So what is on the other side?

So far, it has not proven to be a great place for any kind of tech unless there is a razors-and-blades component. HP went past the wall on printers because it sold ink. Per ounce, the ink in an OEM printer cartridge has typically sold for more than the cost of Dom Pérignon. But for most enterprise tech companies, what has been on the other side of the wall up until now? Hell.

We have already talked about some of the trends that could shape life behind the wall in the age of the cloud, things that will be part of the new normal. But the most important new principle, the one that will have the most lasting and profound change is this: No usage, no money. Revenue will drip in dollars per month from the end users' amount of usage, not gush forth from IT departments' expensive up-front acquisition decisions.

The Consumption Gap will no longer be just a customer satisfaction problem—it is about to become next quarter's revenue problem. Running the product playbook just isn't good enough anymore. The cloud and commoditization have seen to that. Successful tech companies need to run a new and different play. We need to make our product's advanced features get used.

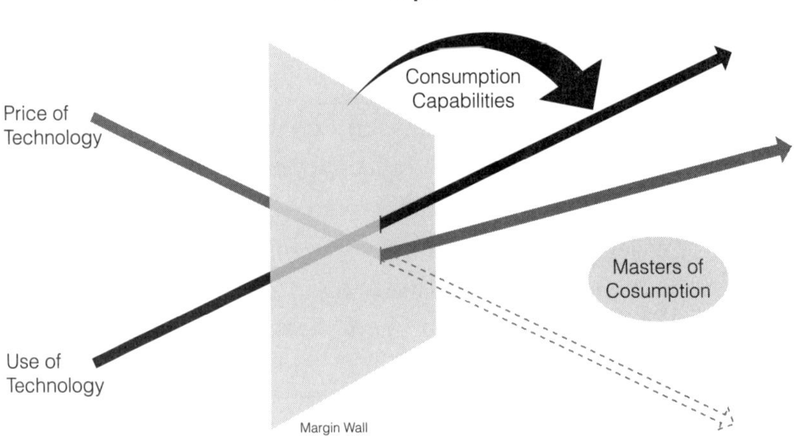

FIGURE 3.3 The Consumption Model

The need for a Consumption Model in the cloud or any pay-for-consumption offering is clear and easy to understand. However, there is an equally compelling reason why tech companies need to think about developing an effective Consumption Model for *all* their products, even their traditional on-premise CapEx ones. If they can keep the value of their products to customers moving up and up—the stickier features, the differentiated capability, the newest innovations—then they can help rescue all of their products from commoditization. Getting to higher levels of product consumption may be the most important lever for turning around sagging product margins and restoring profitable growth that your company is *not* sufficiently focused on today.

The next generation of profitable growth in tech is about growing end-user consumption and profitably selling micro-transactions to enterprise consumers. Building a Consumption Model is partially reengineering and modernizing existing processes, but mainly it is about adding new ones. That is what this book is about. Some of these new ones are extensions of processes already known to e-commerce and financial services companies. Now the consumerization of enterprise tech is coming and all technology companies, consumer and enterprise, need to start becoming masters of consumption. If we continually drive up the value, we will have our best possible chance of defending our price points and defeating commoditization. We can get customers to upgrade to the latest release of our disruptive new products. We can also dramatically increase the return of our R&D investments, which have flattened over the last decade.

The Consumption Models we build will need to be sophisticated. In the cloud, driving consumption of any product's value is not a simple or one-dimensional thought, just like driving revenue growth in the cloud may not be as simple as "per user, per month" math. While that may be the dominant pricing model today among SaaS companies, we believe it is just an initial salvo in what will become the Wild West of usage-based pricing models. As exhibit one, we present Microsoft's summer sale on CRM that we mentioned in the last chapter. We have not even begun to see utility pricing hit enterprise tech. Imagine a pricing model that is:

> *((per user per month fee × number of modules) + premium feature level charges + (number of monthly content subscriptions × price per subscription) + individual content download fees)*

This could be the monthly billing algorithm for a single SaaS user! Remember what we said about the importance of driving a high average price even in the world of micro-transactions? Let's say that, by driving more consumption, we were able to get an end user to move up one level of base product functionality, use one more module beyond the base product, and subscribe to one fee-based data service. By doing this, let's say we raised the average monthly selling price per user by $100. Now multiply that by a large 2,000 end-user business customer. That is $200,000 per month—$2.4 million per year—that could be earned *or missed* by the SaaS company based on the levels of micro-transactions they drive.

You might argue that customers don't want us breathing down their necks trying to drive more consumption. Whether it is a sense of privacy, telemarketer outrage, or George Orwell angst, the fear is definitely a factor in designing the Consumption Model that best fits your business.

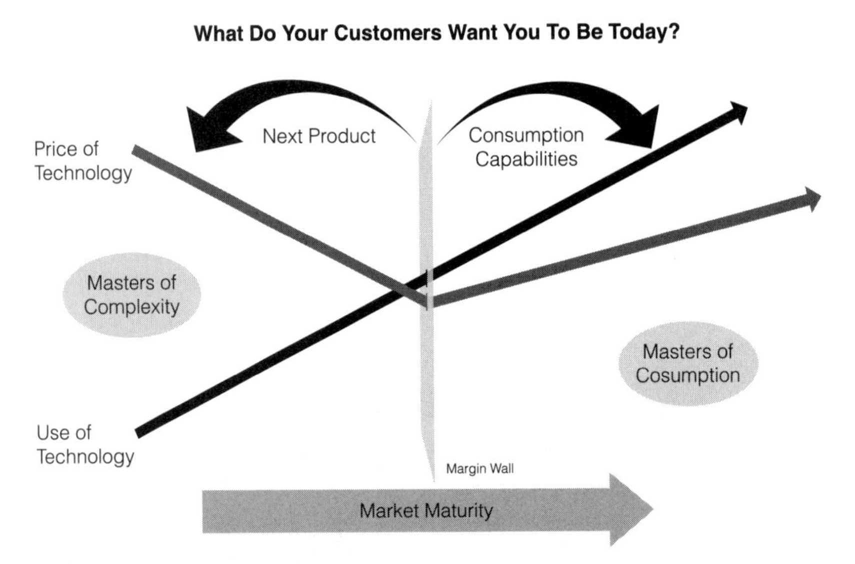

FIGURE 3.4 What Do Your Customers Want You To Be Today?

But we would ask you to step back and think about the messages you are getting from your customers today. What are they asking of you? Do they crave more new features or more support and service assistance? Are they prepared to take on additional complexity or do they want to get it out of their four walls? Do they want you to focus more on the IT department or on the end business users? Do they want more potential capability or more realized business value from the technology they already own? Many corporate customers and individual consumers alike are voting with their pocketbooks. They want help from us. We just need to be smart and as noninvasive as possible in the way we go about designing and executing our Consumption Models.

Now, we are not saying this is true for every company. It is not. At TSIA, we work with nearly 400 of the world's largest technology companies, and no two are the same. Clearly the maturity of your markets is relevant to the urgency of becoming a master of consumption. If you are in a new, bleeding-edge category, then you may not need to concern yourself (yet). On the other hand, if you are in a mature or commoditizing market, read on . . .

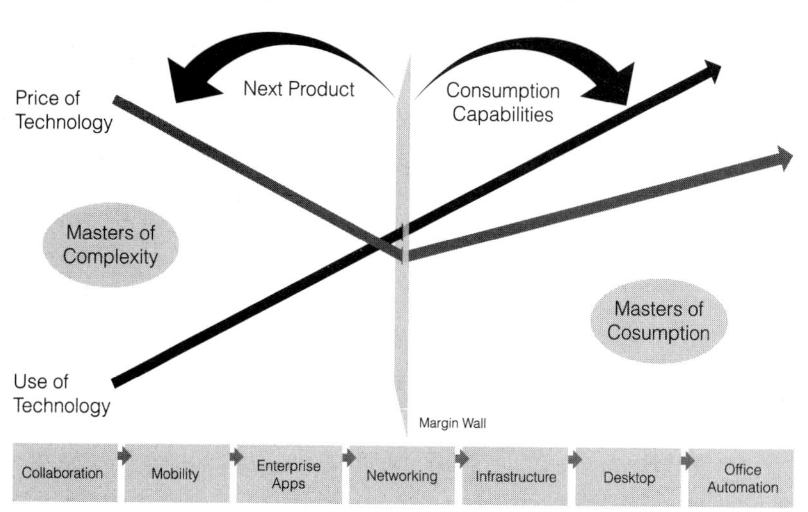

FIGURE 3.5 Market Maturity Drives Demand for Consumption Services

If you have products in small but rapidly growing areas like mobility, there is plenty of greenfield value capture to keep you busy before you run into the Margin Wall. If you are in office automation, you better start becoming a master of consumption ASAP. This is all about natural market forces. All the product categories or market sectors on the right side of the Margin Wall in *Figure 3.5* were on the left side at one time. They too had their days of high prices and high margins. But eventually they hit the wall.

The analogy that keeps coming to mind is about the formation of the Hawaiian Islands. There is a place called the "Hawaiian Hot Spot" that is currently sitting under the Big Island of Hawaii. This island is the youngest in the chain and the furthest to the southeast. As the Pacific plate has moved to the northeast across the Hot Spot's intense heat, volcanic islands have formed. Eventually, that part of the plate moves past the Hot Spot, forever changed, and goes on its way. The island of Kauai is the oldest and the furthest past. The Hot Spot sits there until it finds another weak spot in or around the plate as it moves across and *Boom!*, another volcano is born.

That is how the Margin Wall is. Just sitting out there, waiting to find a weak part of the market to commoditize. Don't believe it can happen to you? I will bet Cisco didn't either, until recently. Price competition and commoditization have come to the switching and routing markets. The high growth and high margins Cisco traditionally enjoyed are under pressure from the likes of HP and Juniper. According to securities firm UBS:

> *Cisco's business model is being challenged and it is pricing aggressively to retain market share. The layoffs (over 6,000 people announced in July 2011) are just a way to expand operating margins since sales growth challenges persist. . .Now the company wants to focus on its core market and is using price as way to preserve it. This is expected to affect stock performance of its competitors.*

Any questions? Has the Margin Wall Hot Spot found another weak spot in the market? You bet. Will the entire category's margins go down as the companies in it use price to vie for share?

You bet. Will leaders try to run a product play to solve the problem with another product? You bet. UBS went on to write:

> *Cisco announced an upgrade to its Catalyst 6500 systems, a move that is expected to help the company preserve its market share by delaying customers from switching to another system.*

It is the play that companies know and love. It has always worked in the past. The Consumption Gap's accelerating effect on the Margin Wall simply changes the rules and makes adding new features a much less effective defense. When your core categories hit the wall, the only option is to dramatically shift your focus to consumption. It is the only way to fight the commoditization and regain momentum. Doing so requires critical thinking around a tough new question: What specific initiatives, if I funded them today, would help a significant portion of my customers better utilize the advanced capabilities they have already bought from me? Which of these initiatives can drive consumption in two quarters? Which will take two years? Those are the tough calls the Margin Wall demands.

In the absence of a successful Consumption Model, most CEOs are forced to react to the Margin Wall using the same old tool: cost cutting.

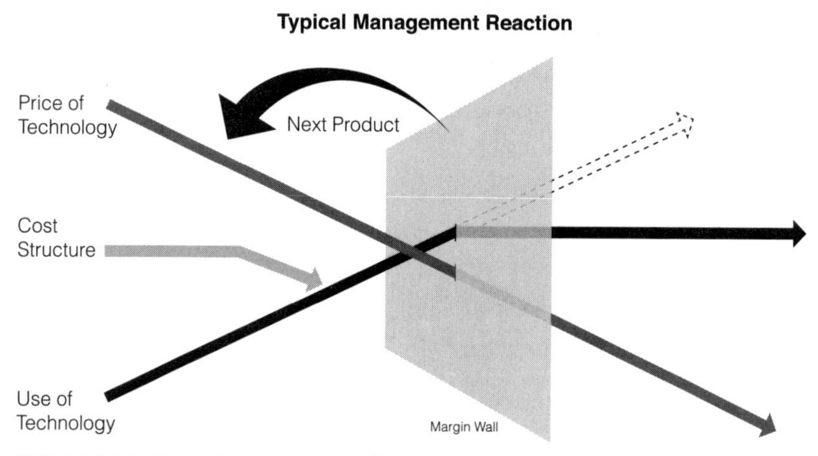

FIGURE 3.6 Typical Management Reaction

You are starting to see this pattern play out at company after company. Enterprise tech company profits may be at an all-time high in 2010/2011, but let's be honest about how they got there. They got there through cost cutting and optimizing the margins of their service businesses. Most of the real top-line product growth over the past five years for large tech companies has been through acquisitions. This has not been a great run for these companies' core products, or their margins. So naturally Wall Street pounces on the CEO because revenues are going up, gross margins are going down, and stock prices are flat.

As we will point out later, there is real danger in the "expense-reduction-as-business-strategy" approach to managing an upcoming Margin Wall, despite Wall Street's demand for action. Sure, take out the cost of dead wood. Become more efficient. That's all great. But we fear too many companies today are going too far in their cost cutting. You cannot cut your way to greatness. We are about to have the opportunity to achieve significant new levels of customer value realization through the Consumption Model. Yet some of these companies get so desperate to hit short-term profit targets that they knowingly cut into muscle. The pressure on the C-suite is immense and career threatening. Together, we have to find a way to work through the transition. We need to come as close as possible to our rosy profit forecasts yet still preserve enough corporate infrastructure and investment to successfully transition to the Consumption Model. There will be more about this in the last chapter.

What else could Cisco do? What other examples are out there of companies that have fought commoditization and won?

Apple has learned to add a second play to its playbook. They have developed a Consumption Model that works for them. They are using software and customer service people to drive the successful consumption of value in product categories once thought to be commodities. They have focused on simplicity,

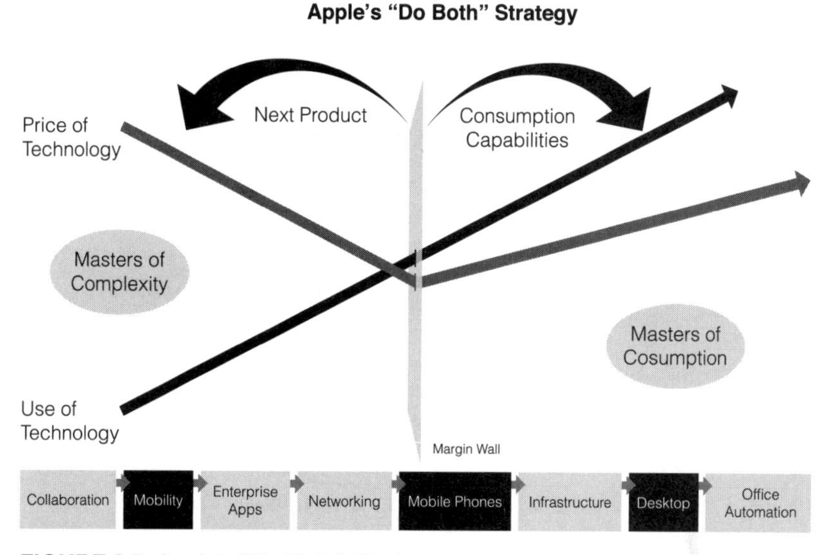

FIGURE 3.7 Apple's "Do Both" Strategy

elegance, and ease of ownership. They opened up retail stores so they could put actual people in service to customers. When every other tech company seemed to be moving their customer support people as far away from the customer as possible, Apple put them in some of the most expensive real estate on the planet. The Genius Bar blue-shirts don't just sell you the phone or the PC, they help you set it up. They train you; they smile when you come to ask them stupid questions. While they have you there, they show you how to do a few more things that you didn't even know you could do. Apple has taken the first steps toward becoming a master of consumption. And in the past three years, they have done pretty darn well.

Here is what is so important. Two of the three main product categories Apple competes in are commodities. But they have managed to take their products and decommoditize them! Now clearly they did not do this solely by closing the Consumption Gap. They did great marketing, designed great products, and innovated with

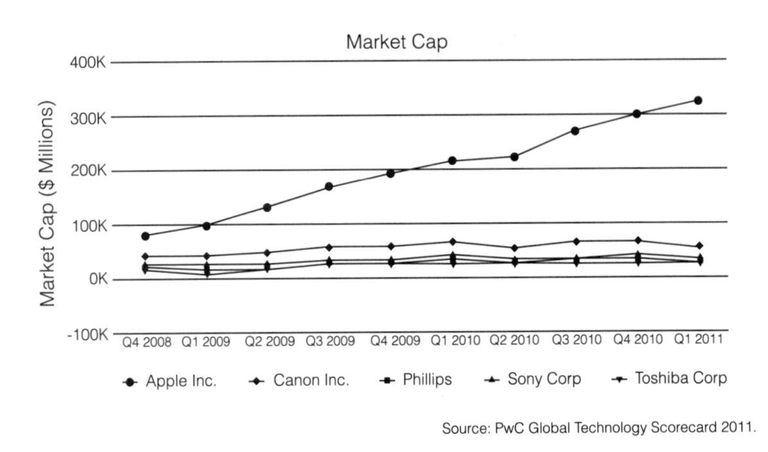

FIGURE 3.8 Market Cap of Major Consumer Electronics Companies

iTunes and the App Store. But what is so great about their laptops that make them worth three times the price of their competition? The MacBook Pro laptops that we have been using to write this book do pretty much exactly the same things as the Lenovo ones we used to use. The Lenovos cost $700 and were decked out. These MacBook Pros cost $1,700 out the door and are missing some key features built into the Lenovo, like a cellular modem. Why buy Apple? Because there is a store near our Silicon Valley office, and we can go over there to get help when we need it.

However, Apple also has another big product, and it is not in an existing commodity market. Apple practically invented the high-growth tablet segment of the mobility market with its iPad. It is a technical hottie and is just at the beginning of its market penetration. Some analysts estimate that Apple will sell over 100 million of them in the first three years in market. That is a $50 to $70 billion market that did not exist in 2009. It was great execution of the classic product playbook. Apple has figured out that they need to be able to run both plays. This is why Apple is valued at 50 percent more than IBM or Microsoft today. A lot of other companies could benefit from that lesson.

Every company that plays in markets near or past the Margin Wall needs to start thinking about a Consumption Model. Even as good as Apple's initial forays are into the world of driving consumption, they are only at the beginning of what's possible. Their model works, but it is costly and hard to scale. Opening stores and putting thousands of people in them to work with customers is not an option for many tech companies. The Consumption Model that the cloud enables will create some very exciting new opportunities for you to get similar results, but to do it at scale and get paid along the way. This is not just a plea to spend more on services. For the first time, we may be able to put the entire company in service to the customer's successful consumption of value. Not just the people in services, but the IP, the hardware, the software, the data . . . all of it. Want to fight commoditization and fly over the Margin Wall? Let's start to innovate not just on what our products can do, but how we can get users to fully use them!

4 Learning to Love Micro-Transactions

THE CLOUD WILL TRIGGER THE NEED FOR A COMPLETELY NEW SET of capabilities from companies that play in the space. It, along with managed or outsourced services, means OpEx budgets are where many tech categories are headed. In an OpEx world, volume matters. That means tech companies need to learn to love micro-transactions (MTs)—to monitor them, drive them, count them, and bill for them. We need to make our ability to proactively drive their consumption a top priority. We need to build an MT revenue gas pedal that we can use to accelerate our growth speed. So what do MTs look like?

Needed:
A High Volume of Micro-Transactions (MTs)

- Per app
- Per user per month
- Per feature level
- Per print or per document
- Per GB data stored
- Per hour of resource used
- Per purchase
- Per data service subscribed
- Per content downloaded

Tens of dollars per transaction...
NOT tens of thousands per transaction.

FIGURE 4.1 Needed: A High Volume of Micro-Transactions (MTs)

We all know these models. We consume them every day in our personal lives. They have become more the norm than the exception. Take a look at your personal bills for a month. Over 70 percent are likely based on usage, less than 30 percent are likely fixed fee or one-time cost. Power bill, cable bill, water, yard crew, cell phone, bank charges? All based on usage. Not flat fee. They are flat fee *plus*. And there is almost never just one base fee level, there are basic, super, and super-duper base fee levels. They charge you at that base amount, and then add on all the other transaction fees that are not covered in it.

This model is no longer just used by consumer utility companies. Tons of sophisticated enterprise products have moved to micro-transaction-based pricing models. You can buy private jet travel by the air hour. Copier companies bill by the page and by the type of image (black-and-white vs. color). So if it makes sense for cable companies, banks, and jet planes, why not tech?

We as tech companies need to learn to love MTs, and so do our corporate customers!

Tech Companies and MTs

All these other industries are busy redefining themselves to be consumed in a service-based offering model. Once they begin to think that way—and bill that way—their strategies for growing customer revenue change.

Tech companies historically viewed both product waves and account development cyclically according to the product playbook: Develop a new product, and then penetrate the market and the individual customers within it. Then build a second product that links to the first one, and repeat the cycle. The faster you are able to get each customer to take on additional products through a new selling cycle, the faster that account grows. As we mentioned before, because product prices were high and big customers needed big systems, the sale contracts were huge. It was a stair-step selling process built on big contracts.

Imagine being a fly on the wall at the tech company headquarters of a traditional product playbook company on the day the CEO first realizes they are at the tipping point. They are going

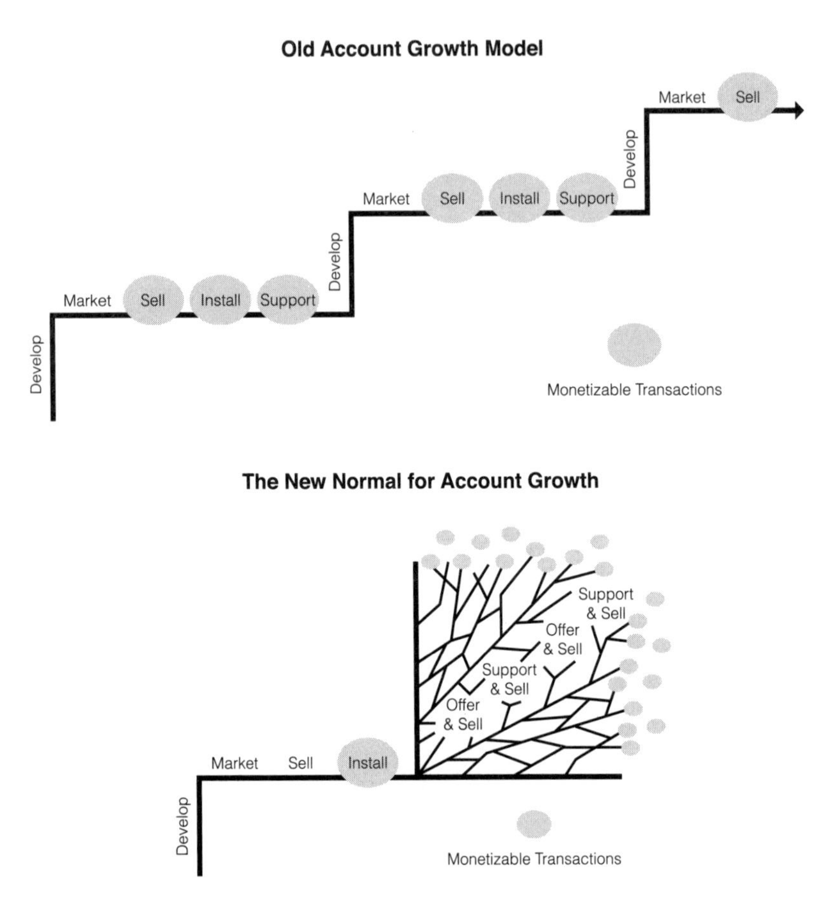

FIGURE 4.2 Account Growth Model Comparison

to miss their number this quarter, not because they didn't sign enough key deals, but because the tiny end users didn't consume as many micro-transactions as they thought they would. Who will the CEO yell at? It's not R&D's fault; the product works. It's not sales' or marketing's fault; they got the IT departments to sign the platform contracts. It's not service's fault; all the customers were up and operational. Who's left to yell at? The answer is no one—and everyone. The question the company needs to ask itself is: Who owns the job of driving consumption?

We used to think the answer was that the service organization should own consumption. More or less, this is probably still logical, which is a case we will make in Chapter Nine. But what we realize now is that driving consumption is a corporate model, not a functional or departmental task. In the new normal, it is a web of micro-transactions, and it will require the involvement of every employee in every department to optimize that web.

At the highest level, it is at least a three-dimensional thought:

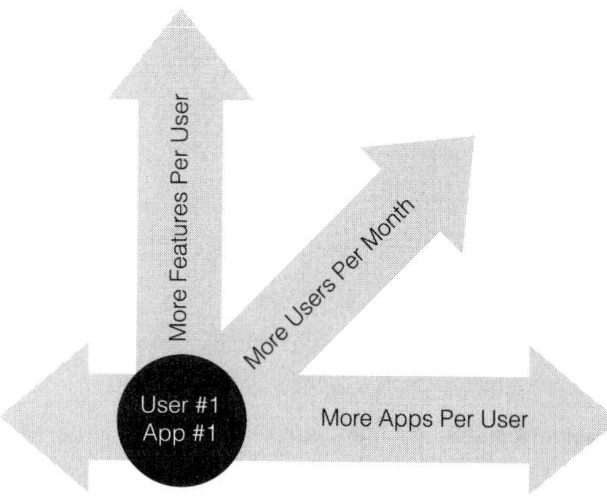

FIGURE 4.3 Three Dimensions of Growing Micro-Transactions

Whether it is end users in a Fortune 500 company, an SMB, a group of high-school friends, or a household, growing the number of MTs is dependent on these three capabilities:

1. **Drive existing application users to consume more advanced features.** *Ms. End User, already good at something? Here is how to go to the next level. Not good at something? Here is how to get good at it.*
2. **Drive new users to adopt an application.** *Mr. Sales Rep in the Northeast U.S. region for Procter & Gamble, did you know you are one of only 15 percent of all P&G sales reps in your region who*

don't have a Hoover's subscription? Mom, did you notice that your daughter is downloading videos to her iPhone? Why aren't you? Let us show you how via iTunes.

3. **Drive existing customers to take on new apps for their group.** *Hey, IT Department and Customer Care Department at P&G, the sales reps are using salesforce.com CRM, why not take on the salesforce.com Service Cloud? Ms. P&G sales rep who has a Hoover's subscription, why not also take on the Hoover's CRM Direct Service? Dad, your whole family is using iTunes, why not get everyone sharing and storing on Apple's iCloud?*

It is likely that these three sales patterns will become important to most tech companies that are tied to micro-transactions. Underperform on any of the three, and you will be leaving money on the table. Of course, you will aim to guarantee as much as possible in the initial platform contracting process. But on top of that there could be huge revenue uplift opportunities as the users adopt the products. These MTs are not a place where the sales force should be spending its time.

As we mentioned, MTs are not just important in cloud models—they are also important in many managed services offers. Since these contracts are also based either on usage or outcomes, driving consumption and business value is a critical and differentiating capability. However, in all these cases, the industry is far from being able to do this consistently and at scale. Let's think through a concrete example of the need to drive micro-transactions and how far we truly are from being able to do it well.

In the office automation business today, managed print or document services is the hot trend. The Fortune 1000 isn't interested in buying copiers anymore. They want to partner with an office solutions company that can take over reducing their print costs, optimize their office workflow, and help them take advantage of new data and document services in the cloud. They want to pay for all that as they consume it. Basically, they are saying to their supplier: "You guys are the office solutions experts! Come in, look at my business, figure out what machines we need and where we need them. Tie them all together whether they are your models or not and maybe even operate the big ones with your own employees. Then look at my internal workflows and business processes and help me manage all that to lower my total print costs and optimize things like cloud backup, search, storage, and retrieval effectiveness."

Welcome to the world of micro-transactions. Each click will drive up the revenue that a Xerox or a Ricoh will get from that account. Each end user matters, each document matters, each copy matters, each upgrade from black-and-white to color matters, and each electronic record matters. So what do Xerox and Ricoh need to do to drive up the total spending of that customer? Let's say the customer is a huge commercial real estate brokerage firm like CB Richard Ellis (CBRE):

1. **Drive existing application users to consume more advanced features.** *Hey, Mr. Executive Assistant, did you know you can copy your original document on this device and print a hard copy on devices in 17 other CBRE offices around the world? Forget paying FedEx bills or sending crummy-looking faxes! Let me show you how.*
2. **Drive new users to adopt an application.** *Hey, Ms. Commercial Real Estate Broker, our research with your peers shows that delivering a full-color book of available properties that meet your commercial clients' unique needs will increase your close rate and decrease your sales cycle time. Here's how to do it.*
3. **Drive existing customers to take on new apps for their group.** *Hey, CBRE IT Department, did you know that we can actually link your copiers to a third-party, cloud-based information service that can allow your users to embed online commercial property listing detail into another cloud service that offers standardized, reusable forms that CBRE can custom design? It will cost you more*

for these additional services, but here's how you can make them save money and increase productivity.

4. **Improve the business processes of the users who link to the devices.** *Hey, Office Manager in the Manhattan office, we believe that your accounting people may not be securing all the confidential documents that they should. Where is the closest secure output device? What if we put a small, secure printer right there in the accounting office along with our new high-security shredder?*

All of these sales patterns are designed to grow a larger account. Remember that every copy and every color counts. Each is worth pennies. The pennies combine to form dollars, and the dollars combine to form profitable growth for Xerox or Ricoh.

But how do tech companies do this at scale and at a profit? Well we sure can't send in a sales rep at these MT price points! We need new, automated ways to sell electronically, and to do it in the context of the end user's workflow.

At TSIA, we believe that cloud offers will adopt the consumer gaming principles of having features unfold as individual end users become more adept with basic functionality. The principles of gaming that can be applied to work applications are becoming well documented. Jane McGonigal, a world-renowned game designer, is perhaps one of the best advocates of how corporations can learn how to engage employees through gaming tactics. An article by Wendy Hawkins outlines some of the key tactics promoted by McGonigal and others:[1]

- Constantly providing positive feedback through points, levels, or performance metrics.
- Providing recognition within the community through rewards or status.
- Creating a community where people can share and compete.

We can see these tactics in play with our own children. When one of our kids gets better at a game, their rank moves up. At the higher rank, the software turns on new features—new terrain, new challenges, new enemies. Imagine your product unfolding as the individual end users "earned" the next level. You would start the end-user experience with basic and easy-to-use core features. Then you would slowly and systematically move every user up the ranks at the fastest pace they are capable of. Get all of your end users up to the highest rank—the most complete level of consumption—and you have defeated commoditization. In consumer gaming, that whole progression might be free, but in the B2B world, it will be "you need it, you buy it."

Imagine your customer's IT department or the business buyer predetermining the features that they need end users of a particular solution to become proficient in to achieve some specific revenue or cost-saving benefit. Then imagine having your product unfold those specific features for individual users in that direction. Imagine having a product that could propose high-value, preprioritized feature adoption offers in intelligent patterns across end users of a global customer? Have a specific set of four new CRM functions that the customer needs the entire sales force to use? Have some key imaging capabilities of your CT scanner that the hospital wants all the techs to be adept at? Imagine being the tech company that could assure the corporate buyer that your offer is the one that can deliver those critical adoption results!

So what is involved in learning to love micro-transactions and being able to drive them in a Consumption Model? As we will discuss in detail throughout the rest of the book, the major *business process* changes will be in the sales, marketing, and customer service processes. But the major *structural* changes that enable those new processes will need to occur within the products themselves. Entire new layers of capability must be added to enable the next generations of customer interactions on a profitable, real-time, high-volume scale.

At a high level of abstraction, creating micro-transactions is about the e-commercization of enterprise technology. We need to apply the best thinking of our top marketing, technical, and service teams to how we can assert some real control over all these teensy, tiny, easily missed revenue opportunities.

This is the next generation of revenue gas pedals. Amazon thinks that way, why shouldn't you?

One other important point about MTs: They evaporate and are not replaceable. Most of us have read about how the airline business works. The core product unit is a seat on a flight on a date at a time. If that flight takes off with a single empty seat, that revenue is lost forever. That unsold seat was a missed opportunity, and the airline will never recover it. The same holds true with MTs. They are incredibly small and easy to overlook, but vital to the overall revenue and profits of tech companies. It is the equivalent of an abandoned shopping cart to Amazon. They may never get another crack at getting that particular revenue from that particular customer. Will they have other opportunities? Sure. But that one purchase? It may be lost forever because they could not get that customer over the goal line when they had the chance.

We need to learn to love MTs—to celebrate their monetization, and weep when they are missed.

Your company may read this and think: "Well that may all be true for us someday, but right now that is just a job for the SaaS providers. As the top layer in the stack, they are the ones who own stickiness, transaction generation, driving volume . . . we are deeper down." Maybe your company makes middleware or network equipment or servers or storage solutions, something that the end user never touches.

Well, it is true that application and content providers (SaaS or otherwise) who are interacting directly with the end users are the top of the micro-transaction pyramid. But guess what? We are *all* in service to the amount of end-user adoption achieved by the stack itself. More MTs not only mean more application consumption, but also more storage, more computing, more data movement, etc. Thus the entire ecosystem of IT providers will all have their revenues largely determined by the success or failure of the end users and their apps. This will show up in two ways.

First is how the giant IaaS aggregators will act. Let's say you are Amazon's Web Services. Clearly Amazon deeply understands that selling computing power and storage is all levered by the volume, nature, and size of end-user activity. For example, Zynga is the company responsible for *FarmVille*, the wildly successful social networking game. When Zynga releases a game, they rent storage and computing capacity from Amazon. The more successful the game, the more Zynga needs to purchase from Amazon. So, Amazon knows they are only one layer down in the MT pyramid from the guys who appear on the user's desktop. Anything they can do to encourage end users to take on more apps, download more content, conduct more transactions, or use more video is going to pay off in more MTs for them. So Amazon is closely studying who the top-of-the-pyramid providers of those offers are. They are already talking to those top-of-the-pyramid providers about how they could cooperate to make their joint end users better consumers of both offers. Could they better integrate the two layers? Could they cross-market a joint offer? Could Amazon build in new functionality that makes it easier to drive more capability or better consumption for the SaaS company?

So now we have the top two layers of the pyramid aligning around a shared goal. Amazon is looking at its layer and trying to help optimize the success of companies one layer up from them so that together they make the total MT pie bigger.

Amazon is also looking down to a layer below them. Let's say that Amazon's CTO has EMC as one of her key suppliers to her monstrous data centers. Since she has the massive purchasing clout that we described earlier, she has already negotiated a phenomenal price for EMC's applications. Maybe it's even an "enterprise usage" license or some other form of all-you-can-eat. So from the early days of the relationship, EMC agreed to lower-than-normal margins in exchange for the huge contract volume that the Amazon deal represents. Everyone seems happy. That is, until Amazon gets dissatisfied with its own margins. Maybe growing competition in an ever-more crowded cloud computing market brings down the prices for Amazon's offers. Maybe Google is entering the market with a big marketing push. We can predict with near certainty that someone will try to undercut them and, when they do, Amazon will want EMC to share the risk and put more skin in the game. Or maybe none of these pressures exist, but Amazon is simply feeling more empowered to renegotiate component supplier contracts as their success grows, just like Walmart does. Either way, the cell phone of EMC's Amazon account executive is going to ring.

Guess what Amazon's CTO wants to talk about? Risk sharing.

As we know, Amazon's offer is based on the "no usage, no money" model. So as they think about their cost structure, it makes less and less sense for them to have fixed-price component supplier contracts against a variable revenue model. Why should they assume all the risks of nonusage? If EMC is truly a partner, they should understand that Amazon also only wants to pay for what they use. So why doesn't EMC get in the game and be willing to base all of its pricing to Amazon on a pay-for-consumption basis? Why not share Amazon's risk? If EMC won't, maybe NetApp will.

So all of a sudden, EMC, a component provider who the end user never directly engages with, is having some of its revenue from Amazon determined by the amount of end-user consumption of Amazon's cloud service. A big part of that is driven by how successful Zynga is at getting end users to adopt the advanced features of *FarmVille*. EMC's growth is now tied to the overall size of the MT pie. So they start to think about what role they could play in driving greater end-user consumption. How could they help Amazon? How could they help Zynga or the other top-of-the-pyramid SaaS or Internet companies?

Now we have three layers of the stack being tied to micro-transactions. Three layers who see that the Consumption Gap is about more than just customer satisfaction. Three layers who are asking the same question: How can we get end users to be more successful?

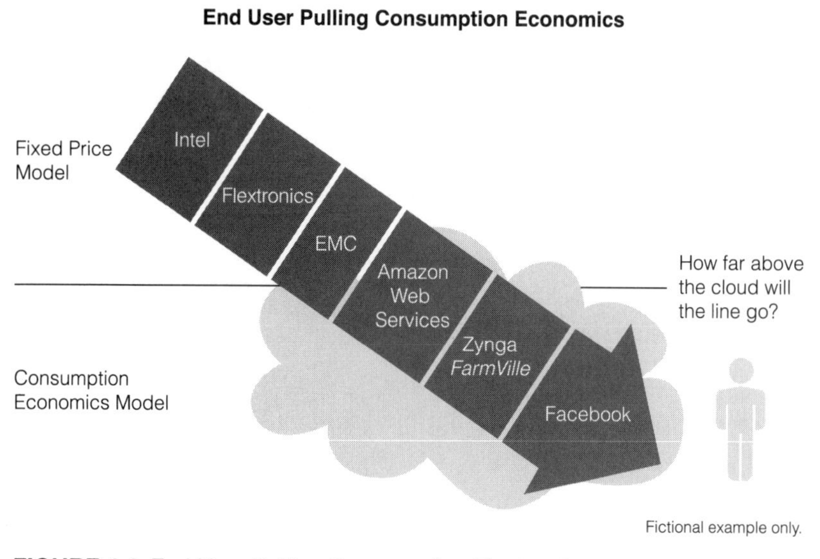

FIGURE 4.4 End User Pulling Consumption Economics

Then EMC's supply chain vice president decides he ought to have a talk with his component suppliers. He wants to talk about risk sharing. So where is the end of the MT consumption chain? Where will the influence of Consumption Economics end?

Everyone who wants to play in the cloud, no matter where they sit in the technology stack today and regardless of who they see as their current customer, needs to have a plan to drive more end-user consumption and MTs in the way they build their products, the relationships they forge, or the services they offer to the layers above them.

The second way that the entire ecosystem of IT providers could have their revenues determined by the success or failure of the end users and their apps is the behavior of the IT department. Grant us, if you will, the notion that virtually every IT department is going to use some combination of traditional on-premise and off-premise cloud-based providers. This means not only two architectures, but two purchasing models. On-premise providers will hang on to the current "pay up front and go onto maintenance" CapEx offer structure for as long as they can. But now they have a new source of competition—not competing against another company's features, performance, or hosting model, but competing against their pricing model. So what does that mean? It means that the IT and purchasing departments may make a decision. They may decide they like the risk shift. They like only paying for what they use. They like aligning their supplier's goals and objectives with their own goals and objectives via an OpEx model. They like taking the shelfware conversation off the table. And they like demonstrating a model to the CFO and CEO where they have reduced the risk of IT investments that don't pay off.

How long will it be before the risk shift hits the traditional on-premise technology suppliers in the form of per-use pricing? As we said, the true innovation of the cloud won't be just about where the gear sits or whether you rent or buy. The true innovation of the cloud will be tying *everyone* (IT departments, application software companies, infrastructure players, component providers, service organizations—both cloud and on-premise) together with the same objective—end-user adoption and consumption of value. And guess what? IT and procurement departments can accomplish this simply by demanding a pay-for-consumption OpEx pricing model.

As we have said repeatedly, tech companies can argue for minimums, we can argue for up-front commits, we can argue for the status quo. But we believe the risk shift is already underway and is just going to gain steam. Whether it displaces 20 percent of your current revenue streams or 100 percent, it is significant. It is a natural evolution of the industry, and few companies will escape it. An executive at a large office automation company estimated that 90 percent of all the major RFPs they are bidding on today are calling for a utility (pay-for-consumption) pricing model.

Figuring out how to play a part in proactively driving micro-transactions will become a core source of shareholder value creation. Optimizing the products, the offers, the marketing, the services—doing all the things it takes to close the Consumption Gap—is about to move onto the front burner at thousands of tech companies. Let's learn to love MTs, because they are worth loving.

Tech Customers and MTs

Tech companies are not the only ones that need to learn to love micro-transactions. Customers do too.

One thing that tech customers have always asked for is predictable costs. Let's face it—no one likes to get a bill for more than they had expected. It's just human nature, whether you are a consumer or a CIO, you always have a preference for predictable costs and long-term visibility. At the same time, tech customers love to complain about the lack of proactive support that they get from their suppliers. They complain about poor adoption by the business user. They struggle to prove IT's value. They also complain about the unused capacity (like shelfware) that they paid for and is sitting idle. What they are really asking for is the supplier to take on more costs to provide additional support. Of course, they don't want to pay that supplier any more than the fixed cost they have already agreed to.

It is the paradox of the existing risk model in tech. The reality is that the customer paid up front and took on the risk of making the investment pay off.

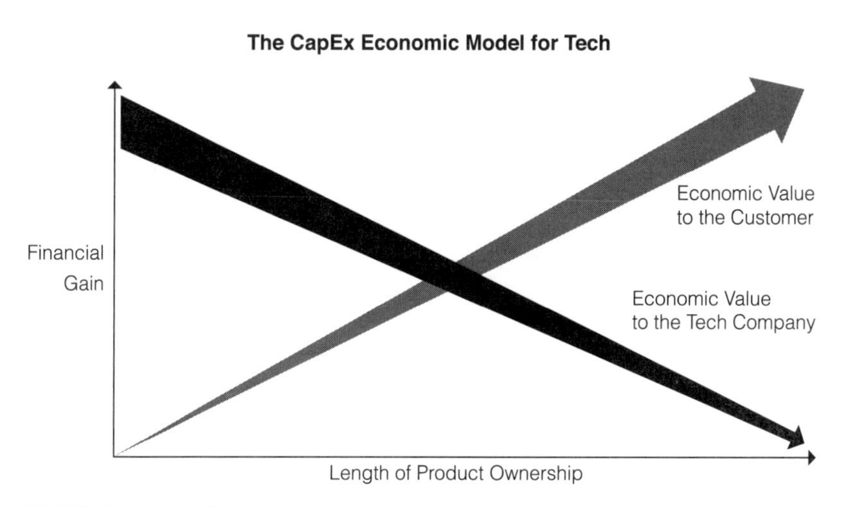

FIGURE 4.5 The CapEx Economic Model for Tech

The tech company took that revenue, recognized it as fast as they could, and is now trying to minimize its ongoing cost of realizing the additional annuity revenue they and the customer have agreed to.

If they provide more proactive support to that customer, which is what every customer wants, their profit margins go down. The classic example is the fixed-price world of support and maintenance contracts. The tech company CFO puts massive pressure on the service delivery executives to lower their costs and improve their margins. These service executives then obsess over how to change people, processes, and technology in a manner that lessens direct customer support—"deflection," it is often called—in order to lower costs. Many times these changes also improve product reliability, but their primary motivation is to lower costs in a fixed-revenue contract.

The bottom line is that the tech customers' demand for fixed, predictable costs is often at the root of their own dissatisfaction over the lack of engagement and support from their suppliers. They have voluntarily signed up for a model where they take on all the risk, and then complain when they are left with it.

The risk shift fixes this critical problem, but *only* if the customer agrees to a new way of thinking.

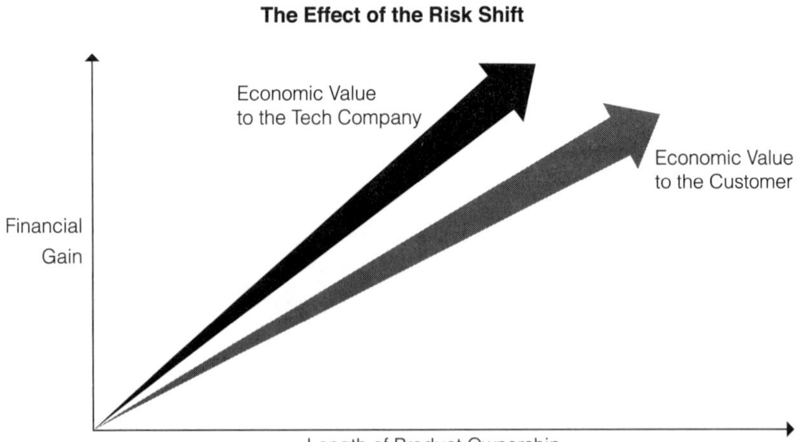

FIGURE 4.6 The Effect of the Risk Shift

The new way of thinking is this: "I will pay nothing up front. Zero, nada. But I do agree to pay for the technology as I use it. The risk will be on my suppliers to actively engage as needed to drive our usage and value from the product. The more usage they drive, the more business value we realize, and the more revenue they get. That means future costs to me that *are not* fixed and are somewhat less predictable." The risk is now perfectly aligned with the customer's goal, but they have to be willing to rethink their expenditure and budgeting models. They, too, have to learn to love MTs.

Now some customers will think to themselves: "Ah, I can get around that. I can negotiate *both* fixed costs and active engagement. I can make my suppliers give me anything because I am such a huge and important customer." Well, this kind of thinking can inadvertently lead to trouble. Take a lesson from dog trainers.

Almost every dog trainer will tell you the same thing: There are no bad pets, just bad owners. What they mean is that most of the poor behavior exhibited by pets can be directly attributed to the behavior of their owners. The owners complain when a dog begs for food at the dinner table, but then they throw it a piece of meat. They complain when the dog won't come when called, but they never formally trained the dog to execute that maneuver. The other thing a dog trainer will tell you is that every dog can be taught to change its behavior at any time. All it takes is a smart training plan and completely consistent and sustained behavior on the part of the owner. The minute the owner stops the new behavior, the dog will revert back to the old way of acting.

So what's the analogy? It's simple. There are no bad tech companies, just bad tech owners. The customers have allowed their suppliers to get away with the old pricing model for the last 40 years. They have voluntarily taken on the risks of getting a return for those investments. They have prioritized fixed, predictable costs over a model that rewards the supplier for delivering increasingly better results. In effect, customers are getting exactly the behavior from their suppliers that they should expect to get, based on their own behavior. So along comes the risk shift with its new pricing model based on low up-front and high, variable usage-driven OpEx pricing. MTs become the currency and get everyone focused on exactly the right thing: end-user success. Customers now have the chance to drive exactly the right behavior from their suppliers. Just like a dog owner who wants to modify his pet's behavior.

Which brings us back to the big IT buyer who is secretly thinking: "Ah, I can get around that. I can negotiate *both* fixed costs and active engagement." Well, guess what? Break the new MT risk/reward agreement even a little bit (insist on caps, all-you-can-eat pricing, enterprise licenses, etc.) and watch your tech supplier immediately revert back to the old model. They will immediately try to cut their costs of supporting you and limit their proactivity once they know the revenue from you is fixed. That is not evil; it is good business, at least in the short term.

Customers should be rejoicing at the risk shift. It is a thing of beauty for them, eliminating many of their most vexing problems, like failed IT projects, inability to demonstrate value, unused capacity, and high IT labor costs. But if they want the new behavior from their suppliers, they have to be 100 percent consistent in their own behavior. Try to outsmart the new model, and the supplier support will stop on a dime. Maybe there are places where that's fine with them. In those instances, they should feel free to negotiate fixed-price agreements. But they shouldn't rest assured that they can have their cake and eat it too. Embrace variable-priced utility models, and their results will go to the moon. Break the agreement, and watch the supplier's old behavior return.

There is another customer behavior that must change. This is a behavior that is and has been central to the IT department's thinking: Prevent the end users from making purchase decisions. The old concept is that "out of control" end users are both underproductive and financially unmanageable. Among other things, the IT department wanted to dictate the hardware and software they used, and even whether or not they could get outside Internet access! Perhaps they thought if they turned on the Internet, all the employees would soon be playing *FarmVille*, consume their hard drives with movies, and download evil software like foreign CRM apps.

It is becoming a more flawed theory by the day. End users have quietly said "no thanks" to a lot of these IT demands. They are purchasing the PCs and smartphones that *they* want. They are putting the apps on those devices that *they* want. They are subscribing to outside services of *their* choice. And you know what? They are becoming more productive. As we mentioned, people's personal productivity is rapidly outpacing their work productivity. They are choosing personal technology that allows them to communicate better, entertain better, create better, and manage their personal lives better than anything that is being offered through the company's IT offerings.

The IT department will need to change its thinking. They are going to have to begin a transition not only to allowing, but embracing, end-user choice and purchasing authority within an overall consumption plan. Yes, they will also still need everyone sitting on SAP or Oracle. But they may need to let people choose their own collaboration tools or data services—that means end users who are empowered, even encouraged, to purchase MTs. Once IT chooses a platform, let the end users buy the add-ons that will make them happy and productive.

This notion will all be predicated on an evolution in procurement and contracting models where the procurement and IT departments consent to allowing end users to make reasonable and targeted purchases without many restrictions (more about this in Chapter Eight). However, since this is really how IT value is increased, we cannot believe that it will take long for them to move away from their addiction to predictable costs and begin to prefer the risk-free model of making their vendors earn the revenue by driving adoption. It just makes sense. If using more of the technology creates more business value (which better be true, or they made a pretty bad purchasing decision), then why not pay as the outcome is achieved and the value is realized. The shelfware problem goes away, the risk shifts to the supplier, and the IT department gets perfect alignment in its ongoing battle to justify its expenses in line with the realization of corporate goals.

Consumption Economics is here and it is going to change the rules for the technology industry. We *all* must learn to love MTs. We need to understand that this is the natural evolution of our maturing technology markets. Tech companies need to get excited by the opportunity it creates to differentiate themselves from their traditional competition. After all, better facilitation of end-user adoption is the basis of a whole new level of value, a whole new differentiation playing field. The smart companies will restructure their products, their offers, and their organization to optimize around the compelling power of the risk shift. They will rush to build their Consumption Model.

And those are the first-mover companies who will win.

5 The Data Piling Up in the Corner

Go out into your garage and look around. See that pile of stuff over there in the corner? You know, that stuff you have been accumulating? It is not stuff that you want to throw away. It has some value. But it is just sitting there. Maybe you even put a tarp over it to keep the dust off. You are legitimately intending to take the tarp off and do something with it, but that day never seems to come. In fact, the pile in the corner just seems to keep getting bigger as you gather more and more stuff. Someday you will do something with it. Someday.

The key enabling capabilities in the age of Consumption Economics are our real-time access to users and the ability to aggregate and analyze usage data. Right now, at most tech companies, that data is piling up on their cloud servers like junk piles up in your garage. Most companies intend, eventually, to do something with that data, but for right now they are just putting a mental tarp over it.

Hewlett and Packard did some pretty great things in their garage in Palo Alto. Now it is our turn. We need to take the tarp off of our pile of data and get to work on it. It represents perhaps the most important new opportunity for this generation of tech. We can leverage real-time user data to change how we develop our products, simplify their use, guide the end users to increased capability and adoption, deploy the best practices in much more targeted and in-depth ways, increase customer value, and grow big, profitable customers. In short, we can change the world by developing a Consumption Model that drives profoundly higher success rates for all manner of technologies. And with that, we can fly over the Margin Wall and rescue our products from commoditization.

Just think about this excerpt from the introduction to *Complexity Avalanche*:

> *Ask yourself a question: What percentage of all the features, of all the technology in the world, are actually being used today? What if we could increase that number by just 10 percent?*
>
> *Worker productivity in developed economies around the world would increase dramatically because they could better use their business tools. Thousands of lives could be saved because doctors and nurses would become more effective at using technology to diagnose and treat disease. Children would learn faster in the classroom and at home through the Internet. The cost of government bureaucracy would go down at the same time its effectiveness improved. People's income would increase as their technology skills got better. The pace of innovation would accelerate. And you'd finally be able to use your home theater remote.*
>
> *The results on the global economy wouldn't be minor, they would be huge.*

That is the power of that pile of data sitting in the corner of your cloud servers gathering dust. Unfortunately, our track record of using customer data is not good. Historically, tech companies have not made accurate, robust customer data a high priority. Heck, some companies don't even have a record of who their customers are or what products they own, usually because they sell through a distribution channel that is very protective of its customers' identities. In the cloud, all that information will be on our servers—every click made by every end user.

In our garage, we need to learn how to apply technology to solving the technology Consumption Gap. We have the tools—probably have had many of them for five years or more—and now we have the data!

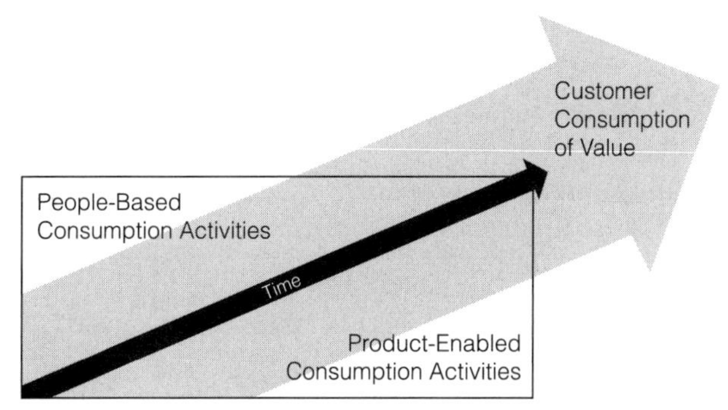

FIGURE 5.1 The Key Play in the Consumption Model

This is the core diagram in the Consumption Model. Virtually every function in the company is going to need to learn how to run this play. Broadly speaking, we need to use our best people and their experiences and insights into how the products should be used and are being used by the most successful customers. Product managers need to begin to identify how they want the product's use to unfold to create an optimized end-user experience and to get full adoption of the product's stickiest, profitable features. The service organization needs to document what it learns about the successes of actual customers and the roadblocks that prevent others from achieving them. In essence, we need to identify the best practices for consuming our product's value. We also need to introduce the ability for the product's consumption to be guided by the priorities of our corporate customers and/or the individual end users themselves.

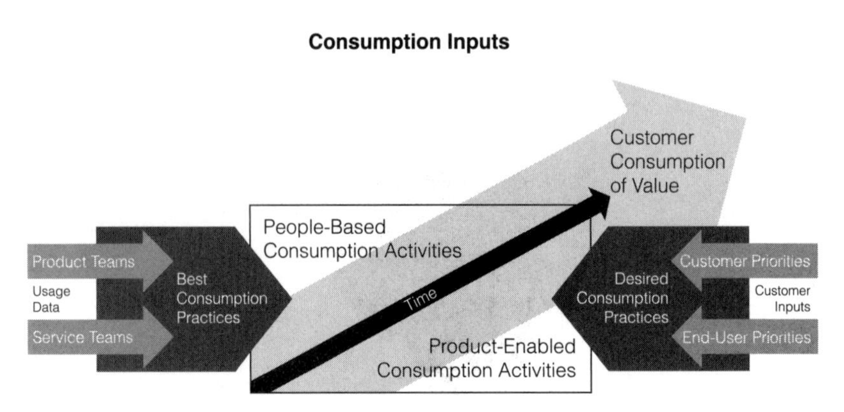

FIGURE 5.2 Consumption Inputs

We need to supplement all these human inputs to the Consumption Model with a new science of usage data analysis. Product managers and service analysts need to apply various BI and analytics tools to big piles of usage data to make sense of what successful users are doing with the products. Once we take all these inputs, both the inputs of the people and the results of the data analysis, we need to form ideal Consumption Roadmaps. There will be many of these, for sure. We will have different ideal Consumption Roadmaps for different vertical markets, applications, and end-user job functions. But think about what all this work will create . . . an entirely new generation of technical IP. The Consumption Model gives tech companies a radical new way to differentiate. It is a powerful new weapon in the war against the evil Consumption Gap and its twin brother, commoditization.

Imagine being able to sell not just all your whizzy features, but also having superior ability to get them into actual use faster and deliver the highest possible value to the customer. This is what Apple offers through its Genius Bar, and it is sure working for them! The Consumption Model is a chance to do the same thing at scale and at an affordable cost, because we will design from the beginning to move much of what we learn into the products.

How do we do that? Well, we need to re-mission some of our R&D resources. Are you sitting down? We may need to take a year or two off from building yet another set of poorly adopted features. We may need to redirect our attentions to building a capability into the product that allows us to easily tailor the user interface for an individual end user in real time based on the ideal Consumption Roadmaps that our best and brightest people (both at our company and at our customers) have identified.

We can't scale all the smart people, but we can scale their insights by embedding them into the products themselves. In the future, we must build a layer of capability into the products that is designed to take the learning that comes from the product's experience in the market and dynamically, purposefully, alters how the product presents itself to different end users in real time.

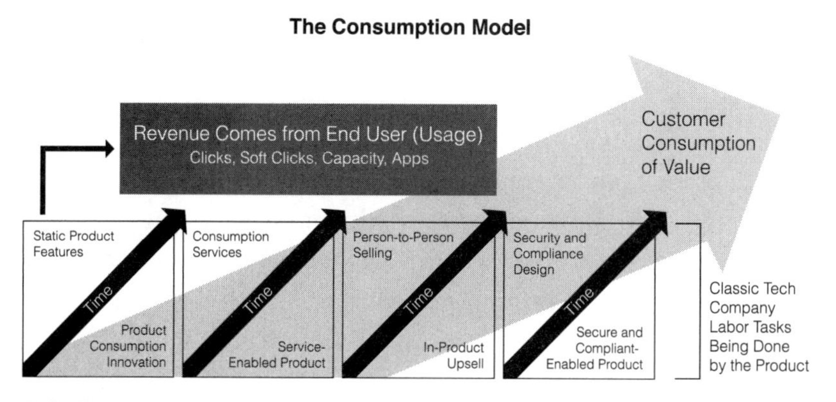

FIGURE 5.3 The Consumption Model

In the old normal model, R&D was about features, marketing was about promoting products to prospective decision makers, sales was about getting big deals, and service was about implementing and fixing broken things. The risk shift and the cloud will change all of that. The data piling up in the corner of our servers and the real-time connection to customers as they use the product will power the Consumption Model in the new normal.

As we said, these are processes from the world of retail e-commerce. They obsess over this stuff. Now it's our turn. We need to make our product and our people smart enough to carry those learnings out in real time and at scale to every end user in the context of their daily product interactions. We can't think of a single customer-facing function at a tech company that doesn't need to adapt to the Consumption Model.

It's funny—everything we have ever read about the consumerization of enterprise technology has been about consumer technologies gaining traction as part of the enterprise stack—BlackBerrys and iPads doing corporate tasks or Google up-leveling its search technology into something of enterprise value. What we haven't heard about is the impact of consumer consumption tendencies on the business model of enterprise tech companies. The freeing power of the cloud will enable more and more customers to try or pilot new technologies. Why? Well, for one, the up-front investment to try things goes to near zero. For many cloud offers, there is no need for installation, etc. Just sign up and go. Secondly, the risk shift means that if they don't like it or can't use it, then they pay little or nothing. All they have lost is a little bit of time. The risk is on the tech company, and their reward only happens if they take the right steps to ensure the customer is successful. This is Consumption Economics.

Taking a small "trial" opportunity and turning it into a huge customer—and doing that around the globe—will become a differentiating capability for tech companies. Companies that survive and prosper on this side of the Margin Wall, where prices are lower and volumes must be high, will need to get really good at this. These Consumption Models will require an entirely new layer of capability in the products themselves, the "e-commerce layer," in order to work effectively. That means lots of in-depth thinking, coding, and partnering for our labs, R&D teams, product managers, and marketing departments.

Service organizations will also get re-missioned to monitor and proactively increase adoption. They too will turn to the e-commerce layer to take their learning and push it out to all users.

In addition, services will become the new high-impact, low-cost sales channel for driving micro-transactions. They will do this both proactively and during the course of customer-originated service requests. This is not "sleazy selling." This will be highly prescriptive and helpful to the customer. Tomorrow's service agents will be armed with unprecedented visibility into the individual customer's situation and have insight into what their particular best next step is. Field service reps will move from being lay-off targets to being invaluable revenue generators.

Sales will still perform the vital role of making the "platform sale" to corporate customers. It is going to be highly consultative, especially if your product is meant to displace an incumbent, on-premise product, or your application is one that the customer is not sure how to benefit from. Salespeople will need to be more service-aware, more highly skilled in business-speak, and more provocative. Unfortunately, all that high-powered selling won't result in much revenue. The revenue generation will occur at the micro-transaction level from inside the product as it is consumed. Sales will be realized by "checking out" the end-user consumption that marketing and service activities are guiding and finally escorting that transaction through to financial revenue recognition.

In the world of MT-based revenue, almost all of our current organizational processes, tools, and skill sets will need to be adapted to a new set of rules. All of these functions will get rethought to become more valuable to the end user. We will become consumption-based organizations in a Consumption Economy. Let's take a look at what some of these functional changes could be.

6 Consumption Development: *The Art and Science of Intelligent Listening*

IN A CLOUD WORLD, TRADITIONAL PRODUCT DEVELOPMENT MODELS are starting to feel a bit old school. Call it New Product Introduction, Agile, Waterfall, PACE. No matter the label, it generally starts with a market requirements document (MRD), proceeds to a business case matching promised revenues to development costs, and ends with an e-mail blast to the sales force asking them to start selling.

As we have said, tech companies are the masters of complexity. We love that role. We love to start with an innovative idea and then build a product around it. It sounds simple, but as we all know, it can be very complex. Where do the complexities lie? Everywhere—in achieving the core technical breakthrough, in the actual product development process, in the underlying limitations of component technologies, in the compatibility with third-party products or legacy systems, in the testing, and in the integration tools. But for most companies, the trickiest element is figuring out and building the features. We knew that we needed to master the functional requirements of the product.

Why is that so important? Well, the definition of adequate functionality determines what the minimum releasable product will look like. That determines the time-to-market, which in turn determines competitive advantage, pricing, and margins. The functionality will also be the core of the product's perceived capabilities and benefits. Get it right and you have buzz and demand; get it wrong and the product fails. Finally, the efficient return on development spending is determined by what parts of development yield features that are frequently used and highly valued by the market.

Where do those critical functionality decisions get made? How do we determine what's included and what's excluded when developing the product? We bet it all on the MRD. Well, in the world of the cloud, the MRD is starting to look more like a boat anchor than a stairway to heaven.

The MRD is intended to be a roadmap for the product. Arguably, next to the core technical innovation itself, this has been seen as the most important determinant of a product's success or failure. In it are hundreds or thousands of educated guesses made by very bright people. The collective goal of those inputs to the development process is to accurately anticipate the market. And that point—that precise point about how best to anticipate the market—is what's about to change. Let's be honest—the traditional development model has some major shortcomings:

- **Customers don't wait.** The MRD is often written one to two years before the product actually ships. That is an eternity in a cloud world. Customer priorities change, economic realities shift, and new innovations appear—all seemingly overnight. MRDs written as "desk exercises" have an incredibly short half-life. Engineers generally have their own view of the product they want to build, so all these important market inputs are often given a backseat to the vision of the product team.

- **Competitors don't stand still.** Almost every MRD paints a picture of how our future product will be better than our competitors' current products. In short, they assume that our competitors are stuck and will do nothing over the one to two years it will take us to get our new product to market. This works when markets are slow moving or when competitors are understaffed and underfunded. It leads to major market misses in periods of rapid, parallel innovation by aggressive product companies with great balance sheets.
- **FCS is just the beginning.** Our product development teams are so focused on getting to first customer ship (FCS), that we often underemphasize the real goal of developing the product. We hit FCS, have a big party, and everyone takes the vacation they had previously postponed. Six months later, when the product is at half the revenue that the business case called for, the finger-pointing between engineering, sales, and marketing gets underway in earnest. Every product person was thinking time-to-market, not time-to-market-leadership.
- **It is okay to prioritize.** Part of why it takes so long to get a new product to market is our goal of getting every feature into it on day one. This made sense in the days before the Internet. Enhancing a software product meant sending out a bunch of CDs. Upgrading a hardware product meant rolling a truck to the customer site. Those days are long gone—our development approach has just not caught up. The incremental cost of pushing ongoing innovation out to customers in the cloud is approaching zero.

Taken together, these practices result in a predictable outcome: We massively overshoot what the typical customer will really use in every new product release. We do this because we cannot accurately predict what the market will want one or two years out, and we do not want to bring to market a product that misses the mark.

If you saw the movie *The Social Network*, it would appear that Facebook's development methodology is a far cry from IBM's or SAP's. The founders of Facebook wrote some code, put it out there, and watched to see what the users would do. Then they wrote the next bit of code in that general direction. If the users started going in a second direction, they developed more functionality in that direction too. As Facebook has grown exponentially, the reality does not seem to be that far from the Hollywood version of application development. A post on the blog *FrameThink* describes the process Facebook follows to develop and release code. The post reports that product managers are almost a useless function within Facebook because product specs are so dynamic. In addition, the post reports that arguments about whether or not a feature idea is worth doing generally gets resolved by just spending a week implementing it and then testing it on a sample of users.[1] We think most people agree that this dynamic approach to feature development has worked pretty well for Mark Zuckerberg and company. MRD? We don't think so . . .

Now before you shoot the messenger, keep in mind that there are very practical reasons for adding lots of features. Don Norman, author of the seminal book *The Design of Everyday Things* and a godfather of the principals of user-centered design, shared with us his belief about why we end up with feature overshoot.

1. **Customers ask for them.** "If the product could just do X, Y, or Z, we'd buy it." That also means that for every vertical market, or every horizontal market, we add features. The more markets we want to compete in, the more features we add.

2. **To trump a competitor's latest innovation.** This is a phenomenon as old as business itself: The product with the most innovative features at the lowest price wins.
3. **Engineers want to show they can do it.** Product developers have a deep sense of pride in what they build and love getting public respect from their peers. There is nothing more motivating than seeing "I cannot believe they pulled that off" posts online.

Now if the impact were just a few wasted engineering cycles, it would be one thing. However, all of this added complexity directly inhibits customers' ability to adopt and consume the product. As Norman explains, "Unused features are not just neutral, they are negative. They degrade the interface and slow adoption of the key features." This is equally true for new products (Windows Vista comes to mind) and major upgrades to existing products. Despite all our efforts to be comprehensive, we often don't really get it right. Enterprise software companies are still faced with a deluge of requests for customization because what certain customers really need the product to do is not quite what our engineers envisioned. It is damn tough for an engineer and a product manager in Palo Alto to accurately anticipate what the market will really want two years in advance. That process is precisely how we end up with a new version of an existing product that includes 1,500 new user features.

Even for the customers who can muster the courage and bankroll to take on the added complexity and adopt the new offering, there is a major downstream price to pay. All of those seldom-used features and complexity have to be installed, customized, maintained, backed up, upgraded, and supported over the five to ten years a typical enterprise technology product is in production. The complexity overshoot is the cost that keeps on costing. In short, technical innovations, no matter how brilliant, that do not get used are a huge waste of money for vendors and customers alike.

FIGURE 6.1 No Feature Overshoot

A Path Forward

It's time we get to a better model. We need to steal a page or three from companies that have grown up developing in a cloud-based world—companies built around the Internet, not ones who adjusted to it later. We don't need breakthrough innovation in our old-school development processes. We simply need to get with the new program. That new program is something we call Consumption Development. In this model, the company does less of the design and lets the market, specifically the end users, do more.

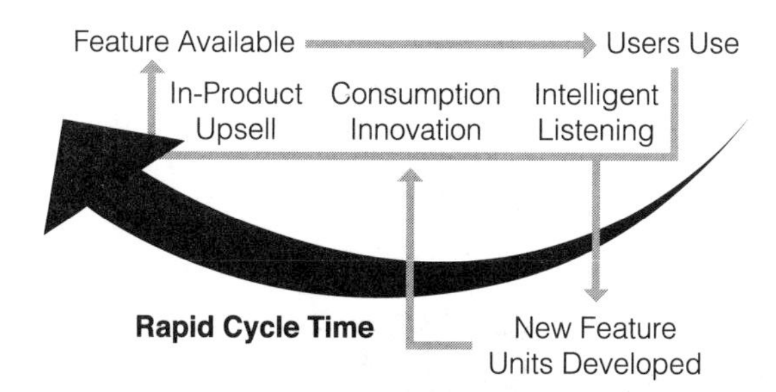

FIGURE 6.2 Consumption Development

Consumption Development has three major elements: Intelligent Listening, Consumption Innovation, and In-Product Upsell. Intelligent Listening takes advantage of user-level adoption and consumption data to predict the features and capabilities most likely to be wanted next. Consumption Innovation rebalances development investments from the traditional speeds and feeds toward entirely new capabilities designed to simplify and guide the end user along predetermined paths to more advanced features, new content, and new value. In-Product Upsell bakes in recommendation engines and offer-management capabilities so that we can tastefully inject consumption suggestions into the end-user experience in real time. In those cases where these offers result in revenue-generating MTs, we need to process those transactions. Each of these is described in more detail below.

Intelligent Listening

For a decade in high-tech, the big have eaten the small (and even some in the middle). EMC, Oracle, HP, and IBM have aggressively been acquiring other technology companies since 2002. But in the cloud, the fast eat the slow. Salesforce.com has demonstrated this in a powerful way. Their stock is up 511 percent over the past five years. That is four times the share price growth of both Oracle and IBM.

To be fast in the cloud world means adopting a new style of development that is based on Intelligent Listening, not MRDs. At its core, this is based on a principle of letting end users speak with their touchscreens, keyboards, and mice. The parts of your solution that generate the greatest actual usage are the parts where your development resources must be disproportionately applied. In the traditional development model, this was nearly impossible. We used customer advisory boards, focus groups, and end-user surveys (as proxies for actual user behavior), which we incorporated into our MRD process. In the cloud, all of the data that enables Intelligent Listening is not just available, it is sitting on our own servers.

Traditional development using MRDs was like a space shuttle launch. We tried to anticipate every need and everything that could go wrong in advance. Once the product was in the market, we had done all we could do until we could get the shuttle back in the hangar for a year or two of repairs and upgrades.

Consumption Development will be much more like the sport of curling. As every good Canadian knows, for every stone thrown in a curling match, there is a clear plan for where the team wants it to end up. While the curler's throw sets the stone's general direction, momentum, and spin, much of the outcome for a given stone is actually dictated by the sweepers. Their light or frantic scrubbing of the ice in front of the moving stone dictates both how far it will go and how much it will curve or "curl." The sweeping adjusts any imperfection in the initial throw. This dramatically increases the chance of the stone reaching its target and making the score.

In Consumption Development, Intelligent Listening is the sweeping. We are still going to put the best product into the market with the most momentum within the time frames we have. We are then going to listen intelligently and innovate rapidly as customers adopt the product and place usage stresses on certain functional areas, often in ways we never could have envisioned in the MRD. Intelligent listening to actual customer usage is how we will make sure the products actually hit the mark and achieve market leadership over time.

To do this, we must fully instrument our products. The design needs to incorporate the most sophisticated user tracking we can manage. We need to learn which specific modules, screens, fields, and functions individual end users are actually using. We will discover where some users are getting stuck and taking much longer than other users to complete a task—even simple things like where in the process of using our products they access the help option and what they search for when they do. These insights are like gold. They give us a way to focus more of our development resources on the parts of the offer that matter most.

As we said in the previous chapter, it is time to take the tarp off the data that is piling up on our cloud servers and use it to build (and rebuild, if necessary) our products in near real time. If you don't have the right data, you need to instrument your products to get it. If you have the data but can't make sense of it, you need to build out new analytics capabilities or integrate someone else's tools.

The analysis of this usage data and the ability to use that analysis to drive short-term product development priorities, as well as longer-term architecture or user interface design considerations, will become a defining source of competitive advantage for your company. Think in terms of what the maximum number of "listening posts" is that you can fit into your solution, and then add one more.

User behavioral data is not the only source of Intelligent Listening. We also need to set up important listening posts in social networking and online communities. Leaders in cloud companies must encourage their employees to be active in the online support and discussion communities associated with the technology categories they compete in. Our developers should build reputations as experts within these forums. The peer recognition they get is yet another feedback loop that keeps them pushing in the right direction. It creates a virtuous cycle where they are constantly thinking market-in, not company-out.

Consumption Innovation

If Intelligent Listening is working, you now have a clear, fact-based understanding of how end users are adopting your product's current features. You are using this feedback to smartly develop new capabilities in small units of work and putting them into your cloud product just as end-user demand is surfacing.

Consumption Innovation takes this to the next step. It involves layering into your product the capability to simplify and guide the end user along predetermined paths to your more advanced features, new content, and new value.

Designing in a tiered user interface is job number one. This is more than just having a basic interface for entry-level users and an expert interface for your biggest fans. It is about building a command-and-control capability inside the product that enables the presentation of features, capabilities, and complexities to individual end users or groups. That is, it allows two dimensions of customization in how users actually interact with your solution: user-to-user differences based on their current capabilities, and within a specific user over time according to a preset Consumption Roadmap.

As we discussed earlier in this book, computer gaming companies do something like this today. They build environments where individuals can tailor their experience of the game and where sequentially more sophisticated interfaces reveal themselves over time based on "earned" or "needed" user behavior.

Dynamic, real-time insertion of new feature units based on best and desired Consumption Roadmaps *without* requiring a lengthy R&D release process is the goal. Such processes simply take too much time at a big company, and markets are not waiting anymore.

Here is the innovative platform your engineers need to develop for your company and its customers:

Development Owns the Product-Enabled Consumption Layers

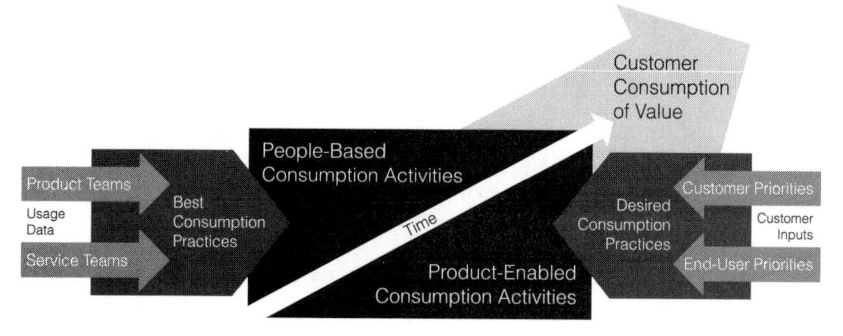

FIGURE 6.3 What Development Owns

The lower half of the center box shown in *Figure 6.3* is what R&D has to build. You are creating the capacity for the user experience in your products to be automated and controlled based on all the inputs from the four outside arrows. These arrows represent people who have something to say about the ideal consumption and utilization path a particular end user should be presented with. This includes your product teams who are analyzing the usage data you got from your Intelligent Listening programs, the service and support teams who are collecting knowledge from their customer interactions about what is and is not working about the product today, the corporate customers who have weighed in about what features and capabilities they deem most important for their employees, and finally, the goals and preferences of the end users.

While the product managers may be the ones creating the Best Consumption Practices document you see on the left side of the illustration, and the sales or professional services departments may be working with the customer directly to create the Desired Consumption Practices document you see on the right side, it is the development department that must build the capability into the products to take these directions and carry them out. All these inputs from all these bright people and all our great Intelligent Listening analysis won't do us any good if the product cannot execute on the learning. That is what Consumption Innovation is all about.

This means that your product may not appear exactly the same to different categories of users. No two refrigerators look the same inside because no two people want to consume exactly the same things in the next few days. Having items in your refrigerator that you don't plan to consume doesn't make sense. When you open the refrigerator door, you want to only see the things you are likely to use. This is the way your product needs to appear to every one of your end users.

The reason you only buy the groceries you know you will need is because if you need other things, you can always go shopping again. Which brings us to the next cool capability in your Consumption Innovation repertoire.

Imagine owning a refrigerator that added the new things you needed every time you opened the door? Let's say you take out the things you bought for dinner and are starting to cook. Then you realize that the recipe calls for an ingredient you missed when you shopped the other day. What if you could just go over, open the refrigerator door again, and the needed ingredient would magically be there?

Your Consumption Innovation engine will not only allow you to have different versions of your product for different end users, it will give you the option to change that product on every user session—even sometimes within the user session. You will be able to predict what features the end user will need next, and turn that feature on through your real-time connection to the customer. Pretty powerful stuff. But you have to build in that command-and-control capability. You have to have a place for the ideal Consumption Roadmaps to be entered. A way to monitor their success. A way to change them. Most importantly, you have to build the product to have a tiered and controllable interface.

Obviously, doing all this Consumption Innovation will need to be some group's day job. Some call it "industrial design," others call it "user interface design" or "user experience management." This is an area where your organization is almost certainly underinvesting. In the past, these groups have been seen as "nice to haves," so when budgets get tight, they shrink. Apple has now shown us that optimizing the user experience trumps pretty much every other engineering goal out there. Be prepared to invest in it.

Online retailers are leading the way here. Not only do they understand abandoned shopping carts and the steps people take before buying vs. wish-listing a product, but e-commerce leaders like Amazon are also figuring out what parts of a page garner the most user interest and how users segment themselves based on behaviors. They use these insights to optimize your shopping experience and maximize the time and money you spend on their site.

In-Product Upsell
The last major lesson to be learned for the future of development comes from the growing number of tech offers that rely on micro-transactions to increase their average customer spending. Again, gaming is a great example. Think *FarmVille* within Facebook. *FarmVille* builds in the potential for users to spend real money as they get deeper into the game and involve their friends in their experience. New attributes, maps, and assets are all monetized over time. Skype, Citrix GoToMeeting, and LinkedIn are all great examples of products today that combine a hugely sticky free or near-free offering with various step-up offers as users get value from what they have built.

This "freemium" business model has proven to be an incredibly effective approach to driving market reach and monetization over short periods of time. Giving away either a time-limited or functionality-limited version of your offering grabs interest broadly and quickly. Upselling drives the business model once they are hooked.

In-Product Upsell is not just a phenomenon limited to Internet applications. Listen to this: A few months ago, one of our TSIA co-workers bought a BMW 335i convertible. It had a 300-horsepower engine. He could have bought the BMW M3 convertible, which is essentially a higher-performance version of the same car with a 414-horsepower engine, but it was an extra twenty thousand dollars. So he bought the less expensive of the two cars, and he loved it, except for one thing: It was not as fast as he wanted it to be. So some aftermarket guy said he could increase the horsepower on the 335i to nearly the same level as the M3—basically give it higher performance even after he took delivery. Our guy agreed and took it to the aftermarket place. The mechanic opened the hood, connected a laptop to the engine, hit keys for about two minutes, and then closed the hood and said, "There you go." And sure enough, it was night and day. Two thousand dollars to throw a software switch!

Selling simplified products at low price points and then monetizing their full capability by just throwing software switches remotely as the customer wants or needs more is the idea. And that idea could be applied to virtually any technology. Imagine a

Ricoh customer service person saying: "Want to be able to print more pages per minute? Sure. You don't need a new box, I can do that remotely for you." "Want two-sided printing? Sure . . . "

Underpinning these thought processes are tons of new product capabilities that need to be layered in. This is the e-commerce of enterprise tech, and once again, the Amazons and Lands' Ends lead the way. Like them, you will need to build into your enterprise technology new layers that monitor and guide the user experience by applying offer management principles. You will need to add recommendation engines: "Users who use this feature also like this one . . ." You will need to build in the capability for in-product, real-time customer service interaction: "We noticed you didn't actually use that feature that you played with—would you like to chat?" You will need to introduce fun, entertaining, short video training apps that your product automatically sends to a struggling user's desktop in a just-in-time model. You will need to build the ability to process MT financial transactions online.

This is not the experience you find in most enterprise software applications or system management tools today. But the cloud-native companies and e-commerce leaders have proven that they work. You need to do the same by treating the end user as your customer, and their usage of your complete solution as a very full shopping cart.

Challenge your marketing and development teams to think about the natural In-Product Upsell points based on the patterns of user behavior you have researched. Design them into the use of the product in a simple way so they do not distract from the overall user experience. Price them in a way that can be purchased on a corporate credit or procurement card. Create a way to "check out" those purchases all the way through the customer's procurement process. And above all, make sure you are able to track abandoned micro-transaction purchases just like Amazon does shopping carts. That is how you will improve In-Product Upsell over time.

Stepping back from the details, this new model has the potential to transform your product development organization and your time-to-market-leadership success. It leverages the unique opportunities of the cloud to simultaneously solve the Consumption Gap your end users are facing and provide multiple paths to revenue for you. Either you will innovate your future products and offers this way, or you will be watching your competitors do it.

What Is the Payback to You?

What could Consumption Development mean? It is moving from old-school R&D inefficiencies to the state of the art in the new normal.

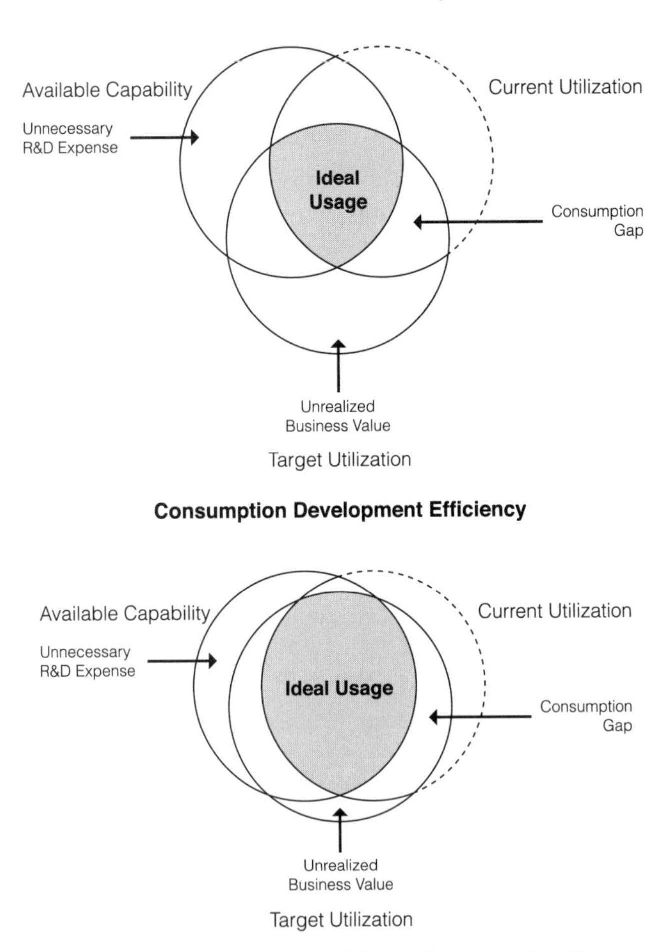

FIGURE 6.4 Moving from Old-School R&D to Consumption-Based Development

Consumption Development removes much of R&D's risk, radically improves its efficiency, and boosts its ROI. It creates better alignment between what the products can do and the value that end users and their companies can actually achieve. Your engineers are *acting* on the market's needs, not trying to *guess* them two years in advance. Your best and brightest people, your company, and your customer's company can now spend their time figuring out how to guide the users to where they want to go efficiently and effectively in the real-time course of the product's use.

Some Spring-Cleaning

So you are all fired up about getting "new school" in your product development approach. There is just one problem. You have double or triple the number of products you need, and keeping them alive is occupying all of your R&D budget. This is the dirty little secret of high-tech: We all like to build new products, and none of us like to retire old ones.

In company after company, we see product organizations where this is the case. To say companies have a "long tail" to their products businesses is a massive understatement. One major software company has well over 90 percent of their revenue coming from just a few product families out of 50. They are not alone.

All these extra low-revenue products are clogging the arteries of your product innovation engine. You want to sprint into the new world, but whenever you do, you get this shortness of breath and tightness in your chest. We get it. We have been there. It is time to do something about it. Think of it as the cholesterol-lowering medication Lipitor for your development, marketing, and sales organizations.

Here are the three steps to future product health:

1. **Snap the 90 percent line.** List your products, by the last 12 months of bookings or revenue, from the biggest seller to the smallest. Now snap a line where the cumulative

bookings or revenue is 90 percent of your total for those past 12 months. The list above the line is what we will call your Horizon 1 products. Here is the kicker: Our bet is that your sales and marketing team could have driven 100 percent or 110 percent of your past year's bookings and revenue off just those products. Think of how much additional market momentum you could have had if you had just concentrated your sales coverage and marketing spending in those few areas rather than "peanut butter" them across the full product list.

2. **Fund Consumption Development.** That long tail of products representing 10 percent or less of your revenues (the ones below the line) likely represents 30 percent to 50 percent of your development team's budget if you were honest about it. If you believe that Consumption Development is your Horizon 2, then it needs to get first dibs on that budget. You need to dispassionately put investments in Intelligent Listening, Consumption Innovation, and In-Product Upsell ahead of all those below-the-line products. Whatever budget is left over, those lower-revenue products can compete for, but you need to be clear that the new Consumption Development projects get funded first.
3. **Clean out the garage.** To make room in your garage for the new Consumption Development activities, you need to deal with all the stuff you have not had the heart to throw out. Now is the time for spring-cleaning. After steps 1 and 2, you have a list of unfunded products and projects. You need to segment them into two piles: "growers" and "shrinkers." Growers are future high-growth businesses that just have not happened yet. These businesses still need to get funded at a level where they can cross the chasm and pay their own way. Shrinkers are everything else. Unless you are Apple with $75 billion in cash, you need to get rid of them. For every shrinker, you need to

put an exit plan in place—end-of-life, outsource, sell, sell-and-agree-to-resell are the usual options.

Now we know this is easy to put on paper and hard to do in practice. The "web of favors" is alive and well. Every product has a mother and father who sponsored them, sold them, and worked on them at some point. Those people (might even be you) will feel like they need to defend those products against this approach. You need an explicit plan to overcome this organizational resistance through communication, education, and change management. If the web of favors is left unchecked, it will keep your development organization forever anchored in the old way of doing things.

Some years ago the United States government had a similar problem with closing military bases. Everyone knew they could and should do it. Many bases were simply not strategic, yet they were extremely costly to keep open. The problem was that every base was inside the state of four or five Congressmen who desperately wanted it to stay open. Each of those Congressmen had other friends inside Congress. It became nearly impossible to get the votes needed to close any one individual base. They got around the problem by putting all of the base closure prospects on one list and made the entire House and Senate vote yes or no on all of them together. It passed and hundreds of bases were closed. That never could have been done one base at a time. This same approach might be the best way to overcome the network of favoritism that allows many individual products to survive far past the point where the economics warrant.

So will Consumption Development become the new standard development process? We think so. But it will take a whole new type of technology innovation. The breakthroughs that can be had are breathtaking. There are new capabilities and insights that could ignite your product revenues, even decommoditize existing products. At your company, Consumption Innovation needs to become just as exciting to your R&D teams as speeds, feeds, and features were in the old school. What will you need?

- Heavily instrumented products and active "listening posts" aimed at real-time insights.
- Behavioral analysts to recognize and analyze patterns of usage.
- Agile development teams and methodologies designed to provide "immediate gratification" to end-user wants and needs.
- Controllable, unfolding products as a design objective.
- Offer management and MT check-out capability.
- Immediate distribution.
- Willingness to experiment, fail, and terminate unused features or whole products.

This is the path to Consumption Development. Your company's journey will be unique, but you should employ these concepts. They will make your products more successful at meeting the requirements for true market leadership.

Oh yeah, and one more thing—you could save millions in development costs as you stop the feature overshoot.

7 Consumption Marketing: *Micro-Marketing and Micro-Buzz*

WE ALL KNOW WHAT MARKETING'S PRIMARY FUNCTIONS ARE. THEY are missioned to create demand and preference for our products in the marketplace. They are supposed to build our brand awareness and create a long shadow that makes us look bigger, badder, and cooler than we really are. If they are successful, they will also lower our sales costs.

The targets of our marketing investments are prospective decision makers and their sphere of influencers. Because these folks are often hard to identify and even harder to target, we are usually forced to wide-cast our marketing investments. These broad campaigns are expensive but, worse, they are inefficient. Maybe they influence some target buyers, but we are also paying to reach thousands of other people that we don't care about. Even new advertising plays such as social media and other "marketing 2.0" counterparts, while they may be more targeted and less costly per eyeball than old-school options such as print, are still inefficient.

So once again we can ask ourselves, why did we put up with this inefficiency? Well, it was the best we could do to reach the decision makers who drove our revenue. Our marketing departments did their best to be targeted, and their absolute best to track the efficiency of their spending.

We believe that the new world of Consumption Economics and the cloud are about to give tech marketing a whole new job and a cool new set of tools to do it with. Why?

- The decision maker for the micro-transactions we need to grow big accounts is now the end user.
- In the cloud, we can know absolutely every one of them.
- Intelligent Listening has given us the usage data from which to construct ideal Consumption Roadmaps. These paths to value will provide unprecedented insight into what these buyers want, need, and are authorized to buy at any point in time.
- We can then put our highly targeted offers right under their noses—guaranteed—100 percent of the time.
- We can track every reaction to every offer—exactly what worked and exactly what didn't.

Marketing's inefficiency problem? Gone. Marketing's "hard to prove the ROI" problem? Gone. Sales asking marketing, "What have you done for me lately?" Gone too.

Micro-Marketing (Finally) Arrives

The next "big thing" in tech marketing will be "micro-marketing"—individual end-user marketing, fueled by dynamic Consumption Roadmaps, targeted specifically at end users based on their unique industry, company, job role, and level of sophistication. Now you might say, "Isn't that the *last* big thing in marketing?" Our point exactly. We have talked about this for years, but have never before had the capability to actually do it like this before! This will be a key new way to grow our revenues in the world of Consumption Economics.

What are we selling with this fancy, sophisticated approach? Features.

Consumption Marketing

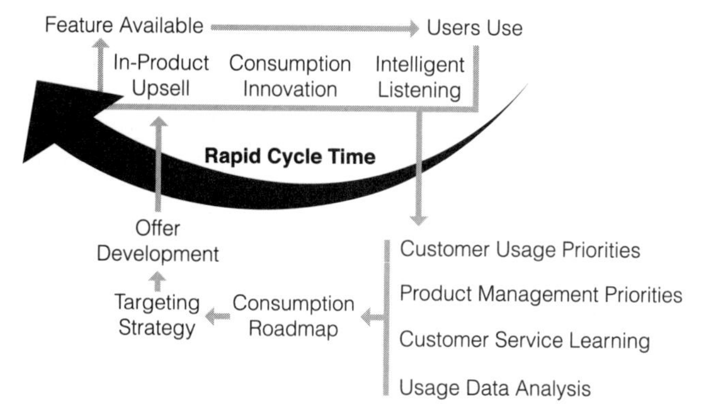

FIGURE 7.1 Consumption Marketing

We have already covered the key enablers of this capability in previous chapters. Micro-marketing is now possible because we can see what end users are doing and we can learn from it. We can learn how our products are being used by banks and how that differs from how they are being used by hospitals. And within a single bank, how does the CIO want the end users to use the product? What modules are most important? What functions are most important? And within individual job roles, what are the power users doing that is different from what the weaker users are doing? How do we elevate everyone to that higher level of adoption and value?

Getting to this level of understanding about our product's journey to high levels of consumption and value is going to become a new, central, and defining capability of tech marketing organizations. It will establish new rules, not just for the next generation of marketing, but also for sales and support.

The second enabler of micro-marketing is our new ability to make offers to end customers inside the actual use of the product, or In-Product Upsell. When you search for a book on Amazon, they not only give you the information you asked for, they also say, "Customers who bought this item also bought _________." This is called "offer management" and is powered by intelligent, real-time recommendation engines. As we said in Chapter Six, your R&D team has to build these and other advanced Consumption Innovation capabilities into your products. While the ability to do this online is relatively new, the basic idea has been around consumer marketing for years. Since the early 1990s, there has been academic research guiding grocery retailers on where to place products on the shelves.[1] Walmart has conducted research using technology that follows the movements of a shopper's eye. They report that most shoppers fail to look at one-third to one-half of the brands on the shelf; shoppers look mostly at the products in the center of the shelf. In fact, shoppers look at the brands positioned in the center of the shelf nine times more than those placed in the corners.[2] This is why retailers put the big-selling or high-margin products at eye level and the peppermint oatmeal on the bottom. Also, they know to put the sauce next to the pasta. Then some consumer packaged goods company that offers both products puts a discount coupon on the two products to direct you from one to the other.

E-commerce sites from Amazon.com to Walmart.com are simply taking these classic retail merchandising principles and doing them online when they make recommendations for additional products based on the products you have browsed or purchased. Well guess what? Here comes the e-commercization of enterprise and consumer technology features.

Imagine an Oracle Apps customer who is in the accounting department doing a basic accounting entry and having a screen pop up that says, "Accounts payable clerks who use this feature also use _________." It allows the basic product to be simple and uncluttered with complex features, and then have it grow in complexity only at the rate that an individual user is ready, willing, and able to take it on. It makes it possible to close the Consumption Gap in a systematic, efficient, and cost-effective manner.

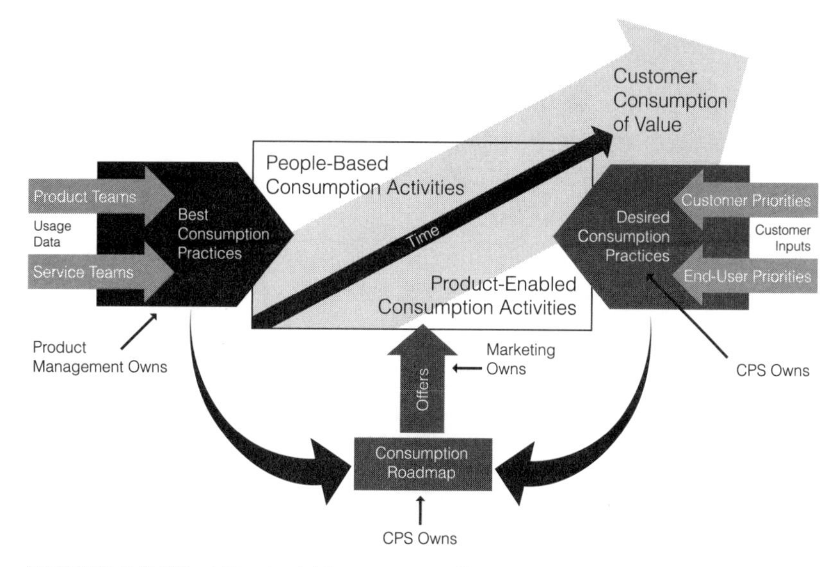

FIGURE 7.2 What Product Management Owns

Underlying our ability to drive micro-transactions by closing the Consumption Gap will be a much stronger commitment to the study of consumption by our product managers. Let's call it "consumption research." The goal is to synthesize all the reams of product usage data we will be collecting, the inputs and learning of our services and development teams, and the product manager's own goals for promoting sticky, differentiating features into a series of Best Consumption Practices. The good news is, it's a concept we should all be very familiar with because it is very akin to something we should all be doing today: marketing research. Here is how Wikipedia defines marketing research:

> *Marketing Research is "the function that links the consumer, customer, and public to the marketer through information—information used to identify and define marketing opportunities and problems; generate, refine, and evaluate marketing actions; monitor marketing performance; and improve understanding of marketing as a process. Marketing research specifies the*

> *information required to address these issues, designs the method for collecting information, manages and implements the data collection process, analyzes the results, and communicates the findings and their implications."*[1]
>
> *Marketing research is the systematic gathering, recording, and analysis of data about issues relating to marketing products and services. The goal of marketing research is to identify and assess how changing elements of the marketing mix impacts customer behavior.*

Okay, how would we adjust this definition if we were to apply it to consumption research? We would not have to do very much. It might look like this:

> ***Consumption** research is the function that links the consumer to the **product** through information—information used to identify and define **utilization** opportunities and problems; generate, refine, and evaluate **learning** actions; monitor **utilization** performance; and improve understanding of **the product's potential** as a process. **Consumption** research specifies the information required to address these issues, designs the method for collecting information, manages and implements the data collection process, analyzes the results, and communicates the findings and their implications.*
>
> ***Consumption** research is the systematic gathering, recording, and analysis of data about issues relating to **using** products and services. The goal of **consumption** research is to identify and assess how changing elements of the marketing mix impacts customer behavior.*

Consumption research shares the same objective as all marketing research: to deeply understand an aspect of the market so that the company can present its offer in an effective, compelling way. The only difference is that consumption research is focused on a narrow dimension of the overall market for a product—the end user. It seeks to answer a single primary question: How can I get an end user to successfully consume the value of my product in the fastest, most comprehensive way? Product managers ultimately should own this IP for their own product. Whether your product management organization sits in marketing or development does not matter. What matters is that the product's internal owner also owns this responsibility.

This body of research—the study of the phenomenon of consumption of a product's value—will not only be central to effective product development, but also to power micro-marketing campaigns. Micro-marketing will require a much greater and more systematic study of the usage patterns around our products. It will also require technology—the Consumption Innovation and Product Upsell layers that we talked about in Chapter Six—so that the product itself can be educated on how to present itself to users according to those patterns.

What are some of the variables that would be in-scope for consumption research? Among other things, we would seek to study and understand the ideal utilization pattern and order of adoption that take place along several dimensions of customer segmentation:

- By vertical industry.
- By job role.
- By typical corporate customer value objectives.
- By typical end-user value objectives.
- By the patterns that optimize purchase cycles.
- By the tech company's competitive advantage and revenue objectives.

These are just the beginning of what could be done. In the consumer world, we might study consumption by age, gender, or other demographic or psychographic criteria. In any case, our consumption research is trying to answer five key questions:

1. How should we segment the end-user population for this product so that we can start off with a manageable number of Best Consumption Practices?
2. Of all the possible capabilities of our product or service, which are most likely to be of high value in each segment? The value priority could be defined by the end user, their company, or our company—most likely a blend of all three.
3. How did the users who are already using those specific features get to that point? What sequence of adoption worked? Items 2 and 3 together will form the basis of our consumption research and result in the Best Consumption Practices.
4. What are the feature progressions and visual stimuli that best guide each segment's users along those paths? We then need to get R&D to turn the other irrelevant and distracting features and functions off for now.
5. What are the user behaviors that will trigger an offer? When and where should we insert them?

Let's pick up on item 2 in a bit more detail. In determining what product capabilities or features are most likely to be of value and why, it's important that we be honest and realize that there are three perspectives to be considered in answering those questions: that of the end user, that of the corporate customer (if applicable), and that of your tech company. While we have said this before, it is important to remember that the objective of micro-marketing is not just about getting the end users to use what they will like best—it's also about making sure that the corporate customer is getting the adoption from their employees on the functionality that is most critical to them. As we will discover in Chapter Nine, the job of extracting and prioritizing the feature use from an individual corporate customer and their end users will likely fall to the professional services team. As part of their implementation exercise they will not only arrive at these Desired Consumption Practices, they will then have the job of reconciling them with the Best Consumption Practices provided by product management to actually build out a tailored Consumption Roadmap for that customer. This Consumption Roadmap will ultimately be carried out by the Consumption Innovation technology built into the product.

All together, this represents a huge new frontier of competitive differentiation. After all, who can really claim to be able to accomplish this today? Nobody. Just think of the value to the customer of such a capability! An ability to basically guarantee the high end-user adoption rates that radically improve the ROI of a technology investment.

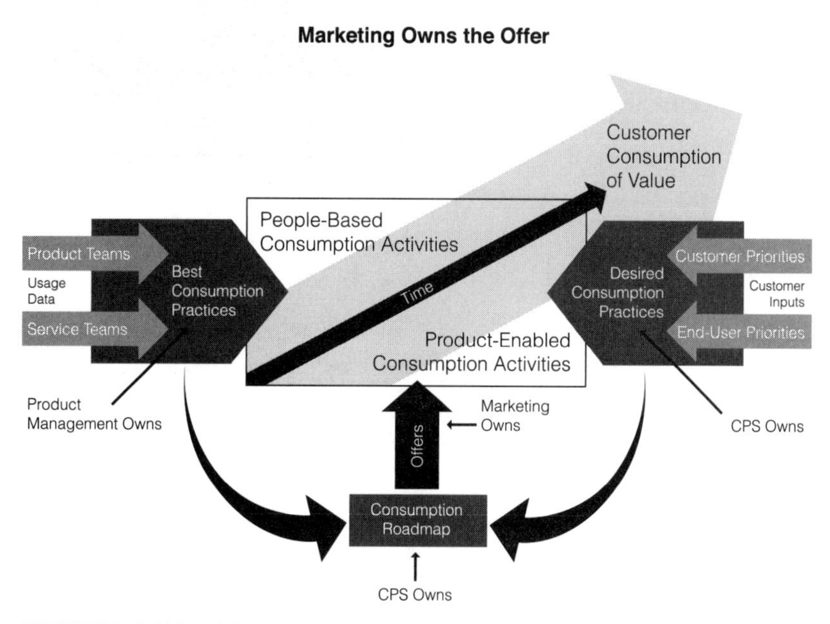

FIGURE 7.3 What Marketing Owns

The next tricky Consumption Marketing task is the creation of the specific offers and making decisions on the frequency and triggers of their use. We will be offering new features and product add-ons (micro-transactions) via offer management within end users' daily product interaction experience. Remember our three dimensions of growing micro-transactions from Chapter Four? Well, we will have offers targeted at all three.

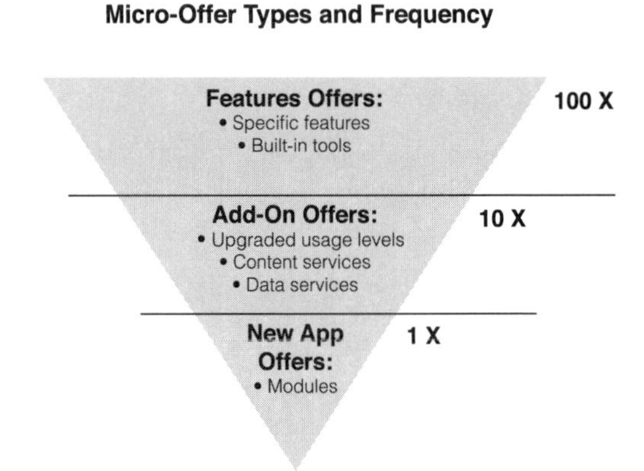

FIGURE 7.4 Micro-Offer Types and Frequency

Our micro-offers need to be helpful, not overly commercial. In fact the most common ones, the feature offers, may not generate any new revenue at all—at least not directly. The most common offers we tee up for the end users have to be quite altruistic. By this we mean that they really have to be designed to further the interests and the priorities for value that the customers and their end users have. Over time, this will result in monetization for the whole cloud/MT pyramid because it will mean more transactions, more data, more processing requirements, etc. But in most cases, ironically, it won't result in immediate gratification to the guys at the top of the pyramid: the SaaS company. That's because few, if any, of us actually monetize individual features. We charge by feature levels. So moving a user up to a single new feature alone usually won't ring the cash register. It will, however, move that user to the next stage of their company's Consumption Roadmap. Over time, as users become more adept and begin to cross major consumption thresholds, this will result in add-on MTs and new application revenue.

This brings us to a critical point, perhaps the most important point in this chapter. There is one more reason why these nonmonetized feature offers are essential: end-user trust. If we are truly being altruistic in our offers, the users will sense that. We can become the trusted advisor to millions of end users all around the world, and not have a single live body in the process. Accomplish this one thing, and you are rich. Why? Because when you do make a commercial offer for an add-on or a new application, they will invariably say yes. The end users will know that you have given them good guidance dozens of other times—guidance that was helpful and smart; guidance that made them more successful and did little or nothing for you as their vendor; guidance that they trust. If a doctor who has gained your personal trust tells you that you need to do three things to feel better, and one of the three involves a purchase (e.g., a drug or a test kit), you don't question it, you just do it. Why? Because you trust their motivations and their knowledge.

Our offers have to be relevant and credible next steps, not huge leaps right to our favorite cash cows. They have to be honestly guided by our research. By this we mean that if the actual usage pattern uncovered by our usage data analysis says one thing but our pocketbook says another, the actual usage pattern wins.

A good rule of thumb for our inverted pyramid offering approach might be the rule of 10:1. In other words, you are going to make ten no-revenue feature offers to an end user for every one add-on revenue offer. And you are only going to make one major offer, like a whole new module or application, for every ten smaller add-on offers. So when you compare the bottommost desirable offer (to buy a whole new app) to the top, altruistic feature offer, it is 100:1. This is how you build end-user trust at scale. This is how you get customers to say yes to your commercial offers. Damage that sacred covenant with your end users, and you will quite simply and quite deservedly get turned off.

The other huge benefit of this approach is that you make the IT department your friend and not your foe. Do you remember when we talked about the discomfort "corporate" will feel as end users drive their own consumption of technology and the micro-transactions that go along with it? Well, if IT is convinced that you have the best interests of the users in mind, you are going to have a much easier time gaining agreement to this variable pricing model.

Micro-Buzz

Here is another cool concept that could be enabled by your new Consumption Model: "buzz" in an account.

We all know about buzz. It is another primarily consumer concept. Twitter has buzz today. So does the iPad. We have also seen it in the B2B world. Early on, it was BlackBerrys with executives and managers. Today, it's virtualization within IT. But all this buzz is about markets in general and leading products within them. The buzz they are creating is global. We could almost call it "macro-buzz."

By contrast, micro-buzz is about getting buzz in a single account. When we watch our kids play online games (okay, maybe it's not *always* the kids . . .), we can see the seeds of micro-buzz. They play games against remote players. Some are close friends, while others are people they may not even know personally. But either way, they get to know each other virtually. They each have an identity, rank, even an aura. They feed off of each other. The power of the group creates energy. It makes them (and us) want to play longer and explore more of the game than we otherwise would. Imagine achieving that same kind of energy among the end users inside one of your corporate accounts.

Now we are not saying that the end users should start taking on alter egos and changing their login names to Darth Vader. But imagine displaying an analysis of how a single user is stacking up against his or her peers in the use of a product. How about having the facility for companies to reward their power users of your product by granting them status and authority to consume higher and higher levels of capabilities? How about making the product so much fun to use that one end user drags his co-workers over to his cube to show them what the product just did.

We shared a silly idea with our good friends at Xerox. You've heard of the hit television show *The Office*? In it, there is an actor named Rainn Wilson who plays the role of Dwight Schrute. Dwight is the slightly sociopathic salesperson and ex-office manager who delightfully lacks both common sense and personal skills. So we asked Xerox these questions: All of your newest production print devices have full-color monitors on them, yes? And they are all connected to the Internet, correct? Well, why not hire Rainn Wilson to make hellishly funny video clips about how to use advanced features of the products and push them out to the device while the machine is busy making prints? It's the same principle followed by car-wash facilities that open up gift shops: "While I have the customer stuck here for 15 minutes, I might as well do some business with them." So while an operator of one of Xerox's machines is waiting for it to crank out a few books, why not do some business with them? Make that experience so much fun that they are running around the office grabbing other potential users to see it too. Heck, maybe they'll even hit the print button again just to see the video one more time! Xerox is not doing this kind of creative feature marketing, but they could. And they should. Every company needs to create micro-buzz through the user experience of their products.

Brian Uzzi, a sociologist at Northwestern University who has helped to pioneer the study of buzz, noted in a *Wall Street Journal* article that the most cost-effective way to generate buzz is to make an exciting product, to create something that people want to talk about. It doesn't matter if it's a critically acclaimed movie or a new piece of technology.

How can you give your end users something about your product that they want to talk about? You have a massive benefit over a movie studio or an Apple app because you already know for a fact that you are going to have your end user's attention. All you have to do is get the offers right. Others have to wide-cast their marketing budget and absorb those massive inefficiencies.

Micro-buzz is all about using your direct connection to the end user to promote ideal consumption offers and news about what is going on with their peer's use of the product—just like games and Facebook do in social groups. Imagine a user being able to see that they are in the top 10 percent of all their peers in getting value for the company through your product? Imagine another user being able to see that they are in the bottom 10 percent? And imagine the company's management being able to see both? Whoa.

There is a lot of precedent for this. Take the Toyota Prius, for example. People who own Prius cars get gas mileage in real life that is very close to the seemingly wildly high estimate shown on the window sticker when they bought it. One reason? It's that cool display in the middle of the dash. People can see in real time how their use of the gas pedal and the brake is either saving them energy or consuming it. This is genius. By displaying that information where the radio station would have been, Toyota has changed not just the car but how people drive the car. We just need to apply this model to high-tech.

If you can get viral adoption of your key features using micro-buzz inside your corporate accounts, you will be not only a favorite vendor to a corporate decision maker, but you also will be a favorite investment for a Wall Street analyst.

Once again, here are the steps:

1. Create ideal Consumption Roadmaps for your customers.
2. Build targeted offers aimed at driving all three categories of micro-transaction: more users per month, more apps per user, and more features per user.
3. Promote and reward end-user consumption success in fun, engaging, and competitive ways.

4. Use the recommendation engine, offer management tools, and In-Product Upsell functionality that R&D has built into the products to make your offers in real time.
5. Track and optimize, add and delete, repeat.

One question you might be asking is how corporate customers will feel about all this intervention you are planning to interject into their end user's daily workflow. It is a very fair question, and one that does not have a known answer. It is just too new of an idea.

We think they can be sold on the concept. Corporate customers not only understand the Consumption Gap, but they understand the power that could be unleashed by better approaches to solving it. We think that—assuming you live by the new rules we just discussed regarding altruism and limited commercialism—they will support and reward the suppliers who embrace Consumption Marketing and all the other tactics in this book because they (the customers) stand to be the real winners.

If you execute your Consumption Model right, then the IT buyer will be able to declare a better return to their business users and to shareholders from their investments in technology. They will also be key players in the process. They will help define the Consumption Roadmap for your solutions and participate in the analysis of actual usage data to course-correct where needed, based on the idiosyncrasies of their company.

In the end, the real beneficiary of employing Consumption Marketing in your cloud-based offers will be your company. In most companies, on most products, the formula for end-user "stickiness" is well known. Pick any of your product managers and ask them if there is a particular set of product features that—once in regular use—create maximum customer satisfaction, utilization, follow-on sales, competitive differentiation, and lock-in. They will answer immediately. Accelerate consumption of those features, and your deal cycles will shorten, renewal rates will rise, and price negotiations will get easier no matter what type of customers you serve.

The benefits to all three parties—you, your corporate customers, and their end users—of getting this right are immense. It simply wasn't possible before now. It takes the unique combination of architecture that the cloud enables. The e-commerce companies understand the power and have mastered the concepts. Regular tech companies never did so because the activity and direct line of sight to real-time usage was not available—they were not directly connected to the end user and all the data was on the customer's servers. Now traditional tech companies can run the same plays as Amazon. But it will take one heck of a lot of work to analyze the data, develop the models, do the marketing enablement of the products, demonstrate restraint in how we message the content, present it effectively (no dancing paperclips, please), and how often we monetize it.

In *Complexity Avalanche,* we made an observation about the dynamics in the phenomenally successful Genius Bars in every Apple Store. Despite all the technology and hoopla, upon close examination you will find that the Geniuses are usually still working one-to-one, and they are working in the context of a specific user's data, goals, and capability. At the end of the day, they are just there to help. And help they do. They help the customers and they help Apple. That is exactly what Consumption Marketing needs to be about: helping customers get the most value from their technology. Do that one thing and the revenue will flow. The micro-transactions will grow, the number of apps will increase, and the customer will spend more. If we lose sight of that primary goal, then customers will rebel against Consumption Marketing techniques. If we turn these into canned commercials—invasive, irrelevant, annoying—then we will simply become the next generation of telemarketers. This is a critical point that successful Consumption Marketers must understand and respect. We have to help, not sell. Helping will sell, but selling won't help. Sorry if we sound preachy here, but we won't get a second chance at this. If the early adopters of Consumption Marketing blow it by being obnoxiously commercialistic, the rest of us may never get our shot.

We think the right tone is the Amazon book referral, i.e., "Customers who viewed this book also viewed this book . . . "—polite, relevant, civil. This simple statement, made at the right place and time, may get your end user to business value faster and more reliably than *any* of our current and expensive models for training and support.

Making Consumption Marketing and micro-buzz work effectively will truly be a company-wide effort. It is but one important part of your company's overall Consumption Model. Excelling at consumption will need to become as much a part of the company's emotional fabric as its core technology prowess. That model must be woven into every design, every business model, and every partner strategy. Consumption Marketing is not a marketing message or a campaign; it is part of the system architecture. Product marketing must design it in, development must build it, services must contribute heavily to the consumption research, marketing must translate the findings into offers, offer management technology must deliver it, services must access it during every service transaction . . . in short, we must become masters of consumption.

Marketing departments will now have the ability to broaden their perspective from surfacing new decision makers to doing that *and* surfacing new end-user adoption opportunities. They can set in motion an automated marketing engine that will never sleep. One that—if it drives just a few more dollars each week from every end user around the world—can drive billions in revenue growth over time. In that process, we can promote the mutual interest of end users, corporate customers, and our own shareholders. No one loses when new value gets created. Everybody wins.

8 Consumption Sales:

After a Great Run, the Classic Model Gets an Overhaul

TODAY'S CLASSIC PRODUCT SALES MODEL IS BECOMING MORE and more challenged each day—yet another victim of complexity.

Surprisingly this model, the one still in use by most global companies, was actually developed by NCR way back in 1887. It was based on the concepts of standardization and mass production popularized during the Industrial Revolution and, later, in the early twentieth century by Ford Motor Company. The CEO of NCR, John Patterson, had the insight that many of these manufacturing concepts could be applied to selling. Essential to the model was the idea that if you built standard products, they would have benefits that were largely the same from customer to customer. These consistent features and benefits could therefore be codified into standard sales pitches and training. Ideal customers could be defined, targeted, and prequalified. Once the sales process was modeled, an idealized profile of the salespeople needed to deliver it could be developed and hired at scale. Because the product, people, and sales process were standardized, the ratio of qualified pitches to closed deals could be accurately estimated, territories could be equally divided, and quotas could be fairly set. In essence, a sales force could be designed, built, and managed much like an automobile mass production line: repeatable, predictable, high quality.

The money was in the products and, like Ford cars, NCR cash registers were pretty standard. Patterson understood how to optimize his business model: Sell a standard product and capture the maximum gross margin. The more we sell, the more efficient the factory that makes them becomes. The more efficient the factory, the better our gross margins. It was a simple, beautiful cycle. Volume equals profit and standardization drives volume.

For nearly 150 years this process worked, not just for Ford and NCR, but for almost every company in the world. Even in modern high-tech markets, this model worked efficiently and effectively, at least until very recently. Companies competed head-to-head with similar sales models optimized to win big product deals and capture the high margins and key customer relationships that came with them.

Complexity is now threatening to render the model obsolete. Over the past decade, almost every major industry has experienced a rapid proliferation of technology-fueled advances. The proliferation is not just seen in the wide variety of tech products in the market, but also in the number of features and functions inside each of them. What's more, individual component products are now routinely integrated and have become interdependent, interwoven nodes in complex networks. Standard products with tons of options are combined with custom applications in infinite configurations, glued together by IT departments and system integrators in ways often unforeseen by any one of the individual component providers. And now, cloud models are creating entire marketplaces of low-cost apps that mean even straightforward businesses like PCs and mobile devices now face the prospect of millions of customers where no two are ever the same. The days of standard products with consistent benefits from customer to customer are coming to an end. Complexity and features proliferation have seen to that. But most companies' business processes, especially sales processes, are built in the classic product playbook model.

That model is looking more and more outdated. The cost of sales is rising everywhere. It is taking armies of product, technical, service, and business specialists to prop up the frontline salesperson during the sales cycle. In many enterprise markets, the mingling of products and services is becoming more frequent and the initial proposals more complex. We just can't keep adding sales-force overlays. Customers are sick of eight-legged sales calls, and we can't keep paying sales commissions to 14 people on a single deal.

Here is the bottom line: Tomorrow's sales process will look very different from the NCR "classic." Winning profitable new customers in cloud, managed service, and outsourcing deals will take very different skills and steps. We need consulting skills and service-oriented compensation models. We need salespeople who think on their feet, absorb complexity and uncertainty, and are uncomfortable selling with canned pitches—ones who are business experts much more than sellers of speeds and feeds.

Neither tomorrow's customers nor your salespeople may truly be able to articulate, much less architect, the end benefit they seek at the time they sit down to talk. The customer of the future will be forced to place a bet on a platform of core technologies, add-ons, and services from a trusted provider who hopefully, over time, helps them navigate the complexity to arrive at the best potential benefit.

This will be just as true for consumers as it will be for the Fortune 500. Deciding between Apple and Android is not about the basic phone—it is about picking a platform from which you will choose from among thousands of apps and features. You are not buying a phone; you are buying options on future communication, collaboration, entertaining, and productivity.

Doing business with technology customers is starting to look more like a lab project than a product sales cycle. We will test this and try that. It will be impossible to design the entire solution until it is deployed and the end users have "voted" with their selection and use of functionality. The idea of canned sales pitches that promote standard benefits is becoming less and less useful—sometimes even dangerous—almost an insult to a sophisticated customer. "Sales consultant" may be a more apt description for tomorrow's salesperson. Of course, our bet is that you already have a few rock stars in your sales team who sell like this today. You are going to need a lot more of them.

Big corporate customers are now busy adjusting their vendor selection criteria to fit this growing reality. They might move their interests from "who has the most features?" to "who has the platform and the best Consumption Model?" "Who can we trust to build us what we need over time—nothing more—and then be the best at driving end-user adoption, which we see as key to achieving business value?" Customers, especially the big ones, are beginning to understand this.

This is the new normal sales cycle and it will place incredible new stresses on tech companies. It is becoming more complex and more risky, one in which services are becoming just as important as the products. The skills and processes we need to win these new deals just don't fit the traditional product sales model anymore.

And it is not just the initial selling cycle that will change in the era of Consumption Economics. As we have said repeatedly, growing a new customer into a large customer may include a limited role for traditional selling because it will be driven by thousands of micro-transactions. It's a world of low price points and no real selling cycles—just the pressurized, systematic, "always on" driving of end-user consumption that leads to higher recurring revenue for the vendor and more true business value for the customer. The job of sales is just to set the stage, not to be on it during the show.

The product sales model that NCR pioneered has been a marvel of success. How many other business processes have survived largely intact for 150 years? It's a classic, indeed, and many of the parts are still working fine. The time has simply come for an "overhaul." Waxing the paint and replacing broken parts just aren't enough anymore. The cloud, complexity, and the risk shift will finally force tech companies to rebuild their sales models, to transform them. We need to develop a new profile for our account executives and retool their processes. It is hugely important to get these transformations right. If not, our cost of sales will continue to escalate, even as product prices and margins are declining. This just can't be allowed to happen.

How many of our current sales workforces have the skills required? In 2001, one of the authors was doing research on what percentage of a historically product-centric sales force had the skills to successfully sell project-based services. The answer from most product companies was around 20 percent. So, what percentage of today's sales force has the skills required to sell in the world of Consumer Economics? Maybe 25 percent? What are those numbers for your sales force? Maybe one-third do it today? And another third could be reskilled? After that, there may be some pruning of the tree. But this whole transformation will need to be done without causing major hiccups to next quarter's revenue. Not an easy surgical procedure, but one that must be performed. Sales-force transformation is climbing the CEO priority list at company after company.

So let's try to anticipate the new normal. How might our post-transformation sales force need to operate? What are the critical sales processes and skill sets that will be required after the transformation is complete? The ideal starting point is to go viral in a confined, targeted space like a single corporate customer—viral along all three dimensions of MT growth. Below are three sales steps that we believe are essential for doing well in the Consumption Model.

1. Winning the platform sale.
2. Selling the pay-for-consumption model to IT, procurement, and the business customer.
3. Expanding the platform agreement and consumption volumes by arming the sales force with consumption research.

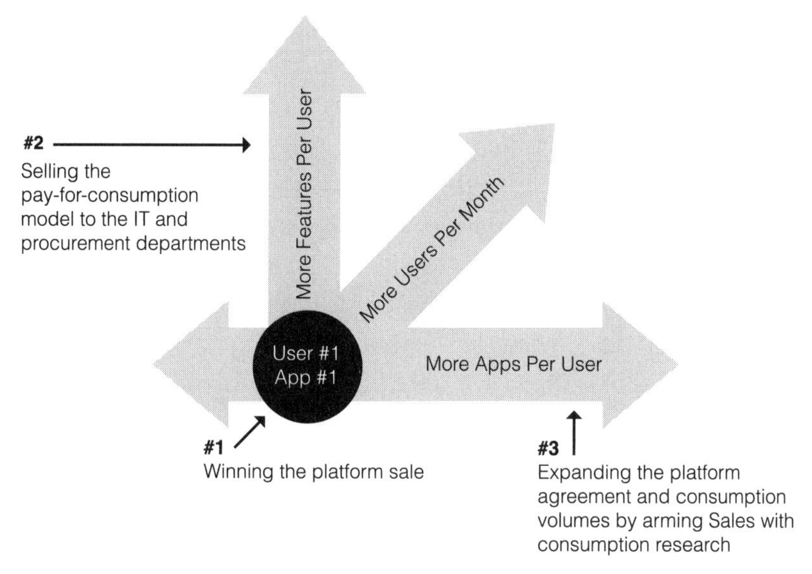

FIGURE 8.1 The Three Critical Sales Tasks

Winning the Platform Sale

The platform sale is the gift that keeps on giving. Every one is an option on a future, growing, and perpetual revenue stream. When Wall Street talks about the quality of a tech company's earnings, they are going to care a lot about how many new platform customer wins the company has achieved.

This is true even though the platform win does not produce much immediate revenue. These deals are very similar to the "design wins" we see in some other industries. Intel works with Apple for months (or years) on how its new chip will power amazing new functionality for Apple computers. At the point where Apple chooses Intel over AMD, Intel gets little current revenue but is almost guaranteed massive future revenue. That's the blessing and the curse of design win deals. It is the future world for lots of tech companies.

So we know these platform deals are important from the vendor's perspective. What about the customer? Well, they are incredibly important to them too. When customers buy into a vendor's platform, they are buying not just the current products, modules, features, and offerings—they are buying into the future Consumption Model breakthroughs that vendor will deliver over five to ten years. They are buying into that vendor's approach to Consumption Roadmaps that enable the tailoring of specific end users, both for their unique job role, expertise, and preferences as well as their personal development over time.

The current state of the art for selling in high-tech has not changed in 20 years. Many call it "solution selling." We have trained our sales teams to listen for specific triggers in how customers describe their issues and pull out a standardized "solution" from their bag of product offers. Sounds a lot like the NCR model.

This works fine when customers already know what they want to buy and from whom. But that's not really what we are paying salespeople to do. If customers already know what they want, they can just order it off of our website or get it from one of our channel partners.

The root of the problem is that customers are now completely immune to this solution-selling technique. We see it everywhere. Discretionary IT budgets are insufficient to meet our growth goals because they are over-committed to legacy systems and projects. The only people who can actually create new budgets for things like our new platforms or managed services are line-of-business executives. Yet every time we try to approach them with our traditional sales model, we find that they actually pay people to keep vendors like us out.

So how do we break the logjam? One way is a breakthrough approach called Provocation-Based Selling (PBS). Philip Lay, Geoffrey Moore, and Todd Hewlin wrote about this in the *Harvard Business Review* in 2009. It is a new sales model to reframe disruptive technology innovations into the business impact a customer can realize with them.

There are four basic elements to Provocation-Based Selling:

1. **The Industry Provocation.** Pragmatist customers tend to reference other pragmatist customers within their industry in making mission-critical technology decisions. To accelerate adoption of disruptive offers, we need to attach them to major industry problems that are unsolved by current technology. The vertical provocation uses the impact of early adopters as a proxy for the industry as a whole, and almost dares the customer not to act.
2. **Referral-Based Marketing.** This includes a series of specific marketing plays that activate all of the referral networks at your company's disposal to facilitate a personal introduction to the line-of-business executive you need to reach. Many of these plays are powered by the new world of social media with its ability for every employee at your company to have "20,000 connections" in their LinkedIn network. Once you get the introduction, the basic approach is, "We would like to come and share some insights around how your peers are solving [*insert major industry problem here*]. Would you be able to free up an hour to see us?"
3. **Provocation Briefings.** The first meeting is critical. If it goes well, the platform sale is yours. If not, you will retreat with dignity and retry at a future date. The difference between the two outcomes is how much the discussion focused on the

customer's business problem vs. how much it focused on the speeds and feeds of your platform. You are shooting for 80/20. Traditional solution selling is more like 20/80. The tipping point in this first meeting is getting an emotive response. At some point, the executive needs to lean forward in her chair, bang her fist on the table, and say, "I don't believe your platform can do that!" When this happens, you've got them.

4. **The Value Scan.** Once you get the emotive response, you need to ask for permission to work with their team for a couple weeks to assess the specific opportunity for your platform in their organization. You don't usually charge for this. Consider it pre-sales consulting. It is a unique opportunity to work with the executive's team and understand the specific aspects of the industry provocation that are most relevant for them. During the two weeks, you will produce a business case illustrating the long-term value of installing your platform or managed service. That business case is the ammunition the executive will use to shift money around and spend her political capital to get your offer into the company.

Provocation-Based Selling is not cheap, but it puts in play important new tools and processes designed to build big accounts, not little ones. Your vertical or solutions marketing people need to develop real insights into the highest potential industry problems your platform addresses. Your field marketing teams need to take a completely new approach to lead generation. And your salespeople need to front-load their sales cycles with executive selling (instead of trying to close the executive like they have been taught to do). This means you need to carefully prioritize the accounts in which you run PBS sales cycles.

In solution selling, we have taught our salespeople to qualify hard on whether a customer has budget to spend this quarter. If not, they are taught to focus on the account in their patch that does. You might think of this as "drive-by selling." No money, you drive by.

In Provocation-Based Selling for platform sales, we need a very different set of qualification criteria. This is hugely important to understand. You are going to be making a very significant bet on each customer you take on. You will have an expensive sales cycle. You will have a big services investment to get that customer and its end users ready and, as we have said repeatedly, you might have little or no revenue from any of those steps. You are betting on the future. You are betting that this customer will someday become a big consumer of your product's value. If you bet wrong and that customer doesn't consume successfully, you are in trouble.

Simply stated, we need to learn to qualify the customer's ability to consume. We either need to make sure that the conditions for deep consumption are already there, or be confident that we have struck an agreement with the customer to put them in place. We need the equivalent of the FICO score in consumer lending. That score tries to anticipate the future behavior of someone seeking a credit card, mortgage, or car loan. Based on the up-front and ongoing revenue success with your existing customers, you need to distill the five to eight attributes that your most attractive customers share and then qualify your next prospect against them.

This kind of smart, selective selling will be key to keeping your sales costs as a percent of revenue under control. If you spend sales and service dollars on customers who can't or won't consume, your sales costs will look horrible. If you choose wisely, the returns on your sales investment will look great.

These consumption attributes will be the basis for how you qualify the vertical markets and specific accounts within them where PBS should be applied. They will also drive how you allocate sales territories for your future sales force. Given how important vertical insights will be, we would expect to see more sales teams organized by vertical industries instead of by geography. These teams will be able to speak the language of the business customer fluently and directly connect the disruptive innovation of their company's platform back to the most broken, mission-critical business problems that industry is facing.

Selling the Pay-for-Consumption Model to IT, Procurement, and the Business Customer

Turning our new customer into a large customer is going to be dependent on capturing a lot of micro-transaction revenue. There are a lot of new things that have to happen to drive that revenue to its optimum level. But there is one primary thing that has to happen. Something totally essential. Something that sales absolutely has to do. We have to get the customer's IT, procurement, and business customers to say it's okay.

Sound simple? Go try it.

Growth along any of these dimensions will ultimately drive more expense for the customer and revenue for the tech product company. IT departments and consumers alike aren't crazy about spending any extra money, especially over their budget assumptions. They like fixed, predictable costs. Unplanned expenses, even highly valuable ones, just do not fit their model.

The salesperson needs to draw a straight line of understanding between consumption and value for them in the selling cycle. Customers complained (often rightly) about spending money on technology that never delivered on its promise. Well, the risk shift removes that problem for them with the no-usage, no-money model. However, it is critical that we educate customers about their part of the bargain. They will be able to buy technology at lower prices and at lower risk than ever before. This is a huge internal win for them. On the other hand, they must be able to accept that vendors who *do* drive usage and value successfully need to be paid.

As we said in Chapter Five, tech companies are not the only ones who need to learn to love micro-transactions. Customers do too.

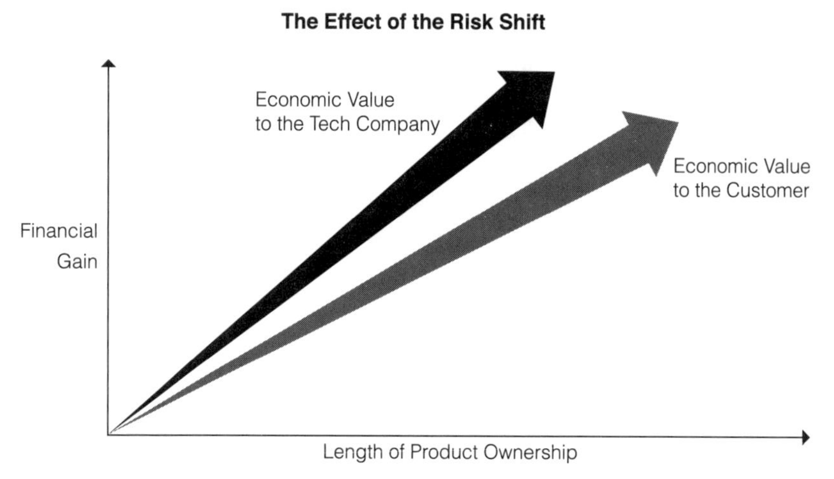

FIGURE 8.2 The Effect of the Risk Shift

We need to get the customer excited about the nature of a new relationship structure where everyone's interests are perfectly aligned. Our sales reps need to articulate the benefits of the risk shift with great alacrity, not hide from the discussion. IT and procurement have to be willing to rethink their expenditure and budgeting models. As we said, if they try to outsmart the new model, they will only hurt themselves. The moment they apply a fixed price to a tech deal, the pressure for the vendor to drive adoption goes right out of the system. They must accept some variability in their technology expenses if they really want adoption. It's that simple. The sales force owns the responsibility for setting that stage.

There are at least three post-sales conditions that must be agreed to up front:

1. Get open-ended sales contracts and relaxed customer approvals of MT purchases by clearly linking end-user consumption to business value.
2. Get the business-side buyers to play an active role in defining their usage priorities so that your Consumption Model can be tasked with driving the desired end-user consumption patterns.

3. Get permission for your service organization to assess and improve customer business processes freely.

This is not just about expectations; this is about business process. We need to work out, and hopefully eliminate, the one-off approval processes for MTs. We also need to work out the book-to-bill process. MTs will happen at light speed. The idea that every MT would need approval is outmoded. We then need to get the businesspeople to give us the consumption patterns that they think are most strategic for key-user profiles. Lastly, we need to get the businesspeople to reveal all their inner workings to our business process experts so that they can make them better companies and better consumers of our product's value. Remember, the Consumption Gap is not just a customer satisfaction problem anymore; it is the key to this quarter's revenue. If the customer's internal business processes or workflows are inhibiting consumption of your product's value, it is now *your* problem. The service folks need to be able to track down and eliminate these roadblocks in order to get the MTs humming.

It is likely that we will drive the customer toward adopting an MT approval and purchasing process that places them into buckets or categories. An example might look like this:

- Critical MTs: Ones that the customer clearly needs, wants, and values. Those should sail through the purchasing process with no approvals necessary.
- Limited MTs: Ones that are deemed helpful but not mission-critical. End users may be assigned a budget for these. An individual end user could purchase MTs in this category up to a point, after which there needs to be some scrutiny by management.
- Exception MTs: Ones that the company deems to be of lower value and might need a process to approve most or all of them.
- Declined MTs: Ones that they simply won't pay for. That is fine; this gets them identified and reduces their priority in your Consumption Roadmap.

What is the one thing you want sales to avoid at all costs? Limits—particularly caps on the Critical MTs. The customer needs to embrace variable-priced utility models. They need to respect the new agreement that underpins the risk shift. Break the agreement, and watch the supplier's old behavior return. The sales force must facilitate this sacred agreement. It is a big change in thinking for the procurement and purchasing departments who don't even want employees buying pencils without permission. It won't be easy, but it is the core sales premise going forward.

It is the job of the sales rep to get everyone on the same play from the consumption playbook, and then stand back and let it fly. The salespeople who can successfully lead that discussion get to stay. The ones who can't, go. This is not about speeds and feeds, price, or discounts. This is about the unique alignment opportunity that the risk shift provides. It is great for everyone, but customers must play ball and must amend their internal processes accordingly.

Expanding the Platform and Consumption Volumes by Arming Sales with Consumption Research

So let's assume that sales did a great job on the first two tasks. We won the platform sale and we structured a true, fair consumption agreement. Now our customer is not only using the product successfully, they are using a lot of it. But there are still boundaries to growth that the sales force needs to deal with. There are still products or features in our portfolio that the customer has not signed up to pay for. There are still MTs that are not happening because the procurement department has failed to see their economic value.

It's time for the sales rep to get back in the picture. It's time for a progress review with the customer. It's also time for sales to expand the platform and up the volumes. We need to set the stage for the next wave of account revenue growth. How are we going to excel at that? We are going to do it with facts. We are going to do it with consumption research.

Remember, we are now the masters of consumption. We have a bigger consumption data set, better analytics, more expertise, more proof, and a perspective that spans multiple industries. We are not salespeople anymore; we are trusted advisors, sales consultants. Our goals and the customer's are aligned. But it is a tricky conversation. The customer might be stubborn about taking on more product capability. They might want to stick with the original plans agreed to in the initial sale. They might think that your product is moving into that dreaded category of good-enough IT.

For the customer, this meeting with the sales rep may make them feel like they are talking to someone who is part leisure travel agent and part Six Sigma consultant. Both parts of the rep want to help get the customer from point A to point B. The Six Sigma consultant is thinking about putting the customer's users on the most efficient, proven paths. The travel agent is thinking about the path less traveled, but where some real gems can be found. These diversions from the basic consumption pattern that the customer initially agreed to may not be obvious to them. They may want to stay on the main highway. But the sales rep is armed with consumption research that suggests that if the customer will allow the end user to journey to some new destinations, those end users are likely to create additional value for the company.

Our consumption research-armed salesperson needs to be both. She needs to be able to show the most efficient paths to value for a particular corporate customer in quarterly or semi-annual utilization reviews. She will do this by having been provided the Best Consumption Models that fit the vertical market the customer is in, the size company they are, and the usage priorities identified by the customer during the sales cycle. She needs to have the real data that benchmarks this particular customer's progress against those best practices as well as along their Desired Consumption Model path. She also needs to show her customer the other sights they should be seeing along the way. Maybe it is an adjacent application or maybe it is a feature set or data service that the procurement guys refused to allow the end users to buy the last time. She needs to expand the scope of the journey that the customer is on with your company. By reaching those agreements to expand the platform, she sets the stage for another wave of MT growth at that customer account.

The Sales-Force Transformation

So if these are the key selling tasks of the future, how does your current sales force map to the required capabilities? Is it time for your company to begin the overhaul of your version of the classic product sales model? As we said, the sales-force transformation is making its way into boardroom meetings and uncomfortable discussions between old friends who have come up through the ranks together. More and more tech executives are realizing that the nature of the customer sales discussion is changing fundamentally and pervasively. They also see that many of their current sales resources are unlikely to excel in the new model.

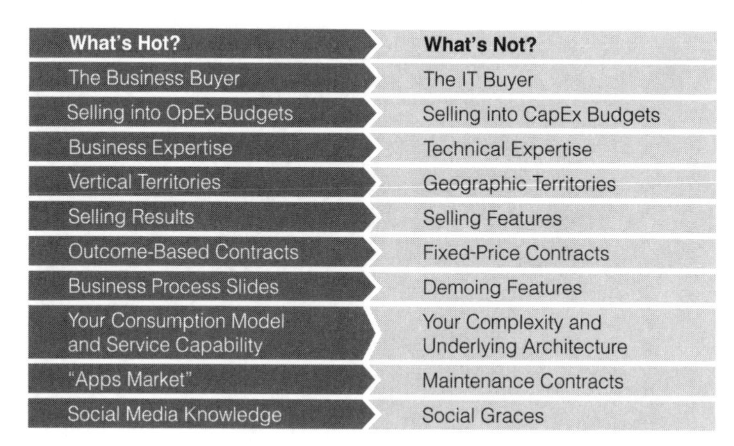

Sales in the New Normal

What's Hot?	What's Not?
The Business Buyer	The IT Buyer
Selling into OpEx Budgets	Selling into CapEx Budgets
Business Expertise	Technical Expertise
Vertical Territories	Geographic Territories
Selling Results	Selling Features
Outcome-Based Contracts	Fixed-Price Contracts
Business Process Slides	Demoing Features
Your Consumption Model and Service Capability	Your Complexity and Underlying Architecture
"Apps Market"	Maintenance Contracts
Social Media Knowledge	Social Graces

FIGURE 8.3 Sales in the New Normal

What has to change in your sales-force transformation project? At TSIA, we partner with PwC's PRTM Management Consulting group when our larger member companies request consulting. PRTM understands this issue. They understand the world of Consumption Economics and how R&D, sales, and service organizations need to transform to flourish in it.

As we work with them on sales transformation projects, it has become clear that virtually every aspect of sales operations needs some updating for the "_______ as a service" or managed services model. Our sales strategies, customer segments, and required capabilities need to change. We need an end-to-end sales process that consistently develops our value proposition in terms of an annuity—one that sells to stakeholders across multiple functional disciplines to get the business group leadership and infrastructure owners to admit and address their inefficiencies. Because account growth will occur by driving lots of little MT dots every day and not the "big deal" dots that our sales guys are used to, the time between buying cycles is now completely disconnected from traditional product cycles. Moving from mega-transactions to micro-transactions and outcome-based reward means a much stronger focus on building long-term account relationships with every aspect of the buyer (economic buyer, decision maker, end users, influencers, etc.). This is a very different mentality than current quarter-driven, transaction-based sales.

Our current commission compensation plans are also not suitable for encouraging this new-normal type of behavior. But complicating our situation is that, if all the revenue becomes MT-based, with virtually every part of the company involved in driving it in that manner, it will be more difficult to align that revenue credit to a sales rep. So how do you compensate them? Do you just pay a reward for the platform sale even though it doesn't drive any revenue? Or do you pay them nothing on that, but give them a piece of all future action as we fill our empty pot with customer gold?

One key modification to our sales thinking in the Consumption Economy is that the sales cycle now extends well beyond the traditional "close" point. So sales and marketing and service better be coordinated in order to continue to "sell" and work with the customer to maximize the value of the relationship. When Tom Siebel first started his CRM company, Siebel Systems, he decided he would not accept anything less than 100 percent customer satisfaction. One of the ways he did this was through sales compensation. Here is an excerpt from an interview Tom Siebel did with *Harvard Business Review* in 2001:

> *The incentive compensation of everyone in the company—the salespeople, the service people, the engineers, the product marketing people, everyone—is based on these same satisfaction scores. So the product marketing people who work on our call-center software receive bonuses based on what our customers tell us about the utility of that product and their satisfaction with it.* ***For salespeople, the bulk of their incentive compensation is paid only after we know the level of the customer's satisfaction—four quarters after a sales contract is signed.*** *That's different from the way it is typically done in the software business, where salespeople are paid when the customer signs a contract. We practice a conscious form of organizational behaviorism. We structure the compensation plan to drive the kind of behavior we want. As far as I know, we're the only company in the world that does this, and we've found it produces the highest levels of customer satisfaction in the software industry.*[1]

When the sales rep got the sale, his commission was calculated. But when did the sales rep get paid? At shipment? No. At installation? No. Upon payment? No. They got paid when the customer's score on a customer satisfaction survey of their top executives reached a specific, high-level target. That one move changed the mentality of Siebel's whole sales force.

These kinds of sales processes may not be new to the service economy, but they sure are new to a lot of tech product companies!

Stronger vertical knowledge and business focus will also be needed because intimate knowledge of the customer's business model and processes is table stakes for conveying value propositions, encouraging micro-transactions, and delivering outcomes. That vertical emphasis will also force us to look at how we segment customers and match the selling approach to the right account. Our salespeople will need intimate knowledge of how the product platform, apps, and services apply to the customer's setting, the different use cases/usage "personas" needed to convey value, and more importantly, how to help the customer see the ways they can maximize value and outcomes from deeper product usage.

Does this sound like a lot for one salesperson to do? It is. We can make great improvements in the number of these capabilities that are innate within our typical sales rep, but we can only do so through overhauling the model. Can we make sales specialists/overlays go away? No. They will still be important to help close gaps in capabilities for a particular customer sales situation. There will still be team-based selling and partnership with the service organization. But hopefully our new, modern sales force will be able to do more of the job than they can today.

What will get in the way? What forces will be working against this vital transformation? Unfortunately, the forces are many, and they are powerful:

- The pressure from the capital markets to hit the numbers you provided using the old model.
- Senior sales executives who excel in the product playbook and don't yet have their real sense of balance in a services model.
- The tendency to promote the sales rep who hit the numbers in the old model, not the sales rep with the skills for the new one.

- The kingdoms that have been built in your sales regions. If the king or queen does not perceive this transformation to be good for their throne, they will make things very difficult with their subjects. They may put counterproductive spins on everything that corporate is trying to do.
- The shortage of people with the new skills you and your competitors want. Have you tried to hire a solutions architect lately? That is what hiring the next-generation salesperson will be like—hard to find and expensive.
- The salespeople who can do it but need to be reskilled will resist taking the training and taking time out of their selling day to learn a new way of working.

Despite these challenges, it must be done. The trends in the market and our cost of sales are simply heading in opposing directions. Failing to get them in better, more synergistic alignment could not just cause lower earnings, it could sink the ship.

9 Consumption Services: *Will They Someday Own "The Number"?*

If you think the transformations to development, sales, and marketing sound tough, well . . .

Perhaps nowhere is the need to transform more important, more urgent, or more delicate than in the services realm. Why? Well, we have two major corporate problems that will fall into the laps of the people who set service strategy.

The first is that services revenue and profit have become the opiate of the industry. We are hooked, addicted. We, our shareholders, and Wall Street—we're all addicted. On average, services represent over 60 percent of the revenues for mature software companies.[1] Even historically product-centric hardware companies such as EMC, HP, and IBM have dramatically expanded service revenues over the past five years. According to our analysis, we estimate there is roughly $800 billion in profitable services revenue industry-wide.[2]

The problem is that the value propositions underpinning most of this service revenue is in trouble. Why? Because the bulk of service revenue generated by product companies is associated with initial product implementation followed by ongoing support of products installed on a customer site. We might market

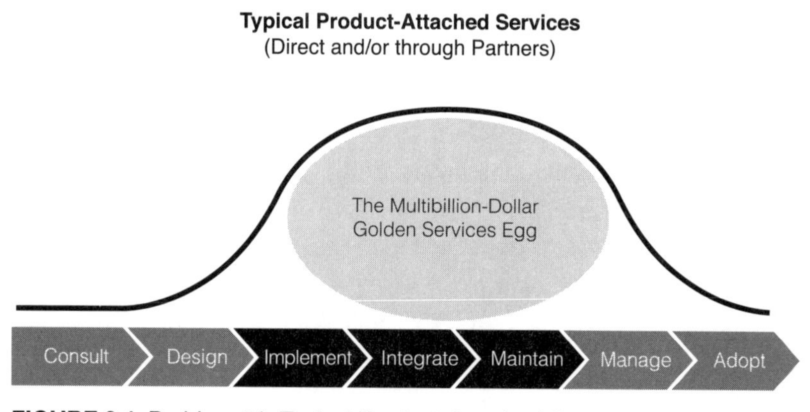

FIGURE 9.1 Problem #1: Typical Product-Attached Services

our company's ability to offer the complete array of services shown in the chart, from consulting to adoption or outsourcing. But the fact is that, except for the very largest product companies that might have a big outsourcing component, the bread and butter of services revenue and margins are in basic technical services like installation, implementation, integration, and maintenance. All of these service functions face an uncertain future. First let's focus on the most profitable of them all: maintenance.

Corporate customers are getting weary of paying 80 percent of their IT budget just to maintain all the complexity we have given them. The complexity that is creating the need for these services is not going away anytime soon, but that doesn't mean that customers have to like where their money is going. As we discussed in previous chapters, customers are increasingly focused on good-enough IT. If they are not upgrading their hardware or software, and the system itself is stable, how much do they really want to pay for traditional maintenance coverage? At the same time, our improved product quality levels and great strides in preventive, proactive support approaches and remote resolution capabilities are also having the unintended effect of reducing support's visibility inside our customer accounts.

Beyond the issue of what price customers are willing to pay for solution availability, hardware and systems customers are replacing older equipment with newer models that cost less and, sometimes, have extended warranty coverage built into the price. The maintenance revenue attached to this new, lower-priced equipment is also going to be less. For things like low-end servers or output devices, it might even go to zero because customers will just run them until they break and then replace them. They are, after all, commodities. This also means more competition from low-cost third parties to service them.

These are all real trends. The intensity of this change in customer sentiment is clear in our TSIA research: Price pressure on maintenance renewals is intensifying and, we believe, will continue to do so. TSIA works hard to provide our members with best practices to mute the impact, but it is hard not to miss the direction of the tide here. Once again, the further to the right your product category is on the Margin Wall chart, the greater the chance this is impacting your current year's—and maybe your current quarter's—service business performance.

We think that there's a tipping point in the total annual service revenues available for a class of technology; that is, all the service revenue that the OEMs, systems integrators, and third-party service providers share for that product category. We call this point "peak service"—the best we are ever going to have it in terms of a growing category of services spending, helping all of us to grow. Once your category moves past peak service, things get much, much tougher. Competitors fight to hold on to precious service margins against declining attach rates, renewal rates, sell prices, and market share.

We believe there are many very big technology categories that are approaching peak service, and that total service revenues in those categories will begin to shrink for the first time ever. This notion of peak service should be enough to get the attention of every member of your C-suite.

But that's not all. Maybe it's not even the most important threat to the Golden Egg of product-attached services. While the trend is undeniable, the time frames are measured in years. We are often asked how long maintenance has as a viable business model at these kinds of operating margins. Wish we could tell everyone an exact date. Once again, it depends on the product category. Older sectors have maintenance businesses that have already begun to shrink. As for the ones that have not hit the wall, like enterprise software or networking, our best guess has been that they still have five years, but not ten. This estimate has been based on these organic maturity dynamics just discussed. But now there is another factor—a wildcard. That wildcard is the cloud.

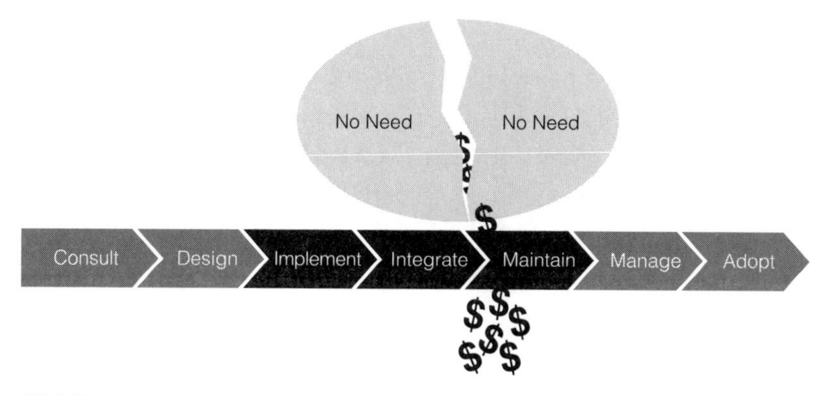

FIGURE 9.2 Problem #1 (continued): Cloud Reduces the Need for Technical Complexity Services

What are the cloud's most well-known selling points? Reduced complexity and reduced cost. There is not a classic install anymore. Maintenance, as a concept, almost goes away completely. Integration and customization tools put more capability in the hands of the customer, so there is less need for them to sign up our service teams to do it. The cloud promises a reduced need for the high-margin technical services that we have become addicted to for generating our top- and bottom-line quarterly performance. Pick an enterprise tech company, take away the profit margins from their services business, and then stand back and take a look. You will not like what you see.

The timing and scale of the cloud's accelerating influence on all of these trends is still difficult to gauge. There are two dimensions to the trend. The obvious one is that every customer who moves part of their IT into the cloud is a customer who may move off the service P&L and onto the product subscription P&L. This makes service profits go down. We will talk more about this surprisingly tenuous shift and what it means if we handle it wrong. The bottom line is that if these customers don't need installation and maintenance, they won't be writing us a check for them. However, the actual revenue loss due to the shifting mix between traditional on-premise and cloud-based IT is just one dimension of the problem. It is going to be what it is going to be. We think it is going to be big and fast because the market is so primed to get rid of the old models and jump on the new ones. The best we can do is to manage those transitions intelligently. TSIA has lots of content on how to do that.

The wildcard is not in the substance; it is in the appearance. The cloud will shine a very bright light on the IT department's spending on services that deal with installed technical complexity. Those expenditures will look archaic. They will get questioned. IT departments will intensify their efforts to reduce them, to bring them into line with what they can get in a cloud model. The biggest short-term impact of the cloud on the Golden Egg of services might not be about losing a service contract customer;

it may be the price concessions required to keep the traditional product customer on maintenance. The brave game of "Chicken" that we play with customers who demand maintenance discounts could get a whole lot more common.

Here is the bottom line of our first service problem: We have billions of dollars in industry profits at stake on a value proposition that is running out of steam. Customers feel like they are held hostage by it, and quite honestly, who can blame them? It is our products that are complex, hard to use, hard to customize, and that require constant attention to keep running. The idea that customers are spending a huge percentage of their budgets to pay us to deal with a problem that we created seems a bit absurd. What we are really selling are insurance policies—insurance against new project failure, insurance against the costs of downtime. Sure, we are also keeping our software developers adding more features to the products, but the Consumption Gap and the awareness of the cost of complexity are making a sizable percentage of customers just sit still. So what good are those new features to customers who don't plan to use them?

We need to switch the value proposition of product-attached service offers while the plane is still in the air. If we don't, the Golden Egg is going down with the plane. No one can let that happen. No one. As we will discuss later, it is even risky to say that the replacement for service annuity revenue and margin are product annuity revenue and margin. That might be a dangerous assumption simply because the price elasticity performance for products and services in tech have been so completely different. History has proven over and over that product prices and margin go down over time, while service prices hold up pretty well. TSIA research proves that service margins can actually go up over time, not down. If we simply swap the services revenue for the product subscription, and then it commoditizes, what do we have left? Once again, take any enterprise tech company today and take away its service profits, and yikes! While the buzz of the cloud may keep product subscription prices up for a while, they will come down over time as those cloud offerings achieve scale and attract competition. If we sacrifice our separate services revenues along the way, we are walking a tightrope without a net.

So what to do about the cracking Golden Egg of product-attached services is problem number one for the strategy folks.

Problem number two is about the gaping hole in our Consumption Model strategy. Even if we are successful in adapting the principles of Consumption Development, Consumption Sales, and Consumption Marketing, we still have a major gap to fill. Remember when we talked about being a fly on the wall when the CEO first realizes that no one in the company has the job of driving customer consumption of micro-transactions? This problem is a tough one, and solving it is going to force us to engage in unnatural organizational acts of extreme proportions.

Simply stated, here is the problem: We currently have no capability in sales or services to effectively help grow the customer in the age of Consumption Economics. The risk shift puts the onus on us to help customers through this pay-as-you-consume evolution, yet we have no group charged with guiding them through it and, in return, capturing the maximum dollar value of their future spending for our company.

Let's think through a typical cloud customer's life cycle in terms of the necessary steps to optimize their total IT spending. An analogy beautifully represents the challenge. It is a simple picture. Mentally resign from your current role for five minutes and become an arborist.

Our company wants to plant a pretty, vibrant, and well-loved tree at the customer's site.

What would our company have to do to successfully accomplish this? The three key steps would be:

- Step One: Obtain permission to dig.
- Step Two: Plant the tree.
- Step Three: Grow the leaves and the branches.

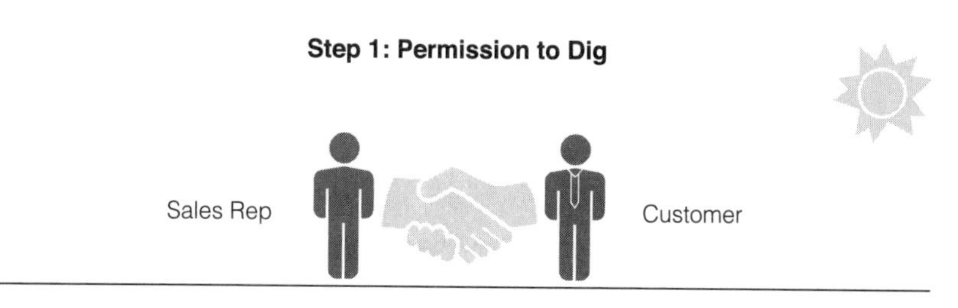

FIGURE 9.3 Obtain Permission to Dig

Obviously, before we can plant the tree, we have to obtain the customer's permission to do so. What is our tree? It is our cloud platform. It is the model, the concept, the capabilities, the apps, the economics, the ecosystem of compatible third-party add-ons or services—all the core selling points that separate our cloud offer from competitive offers, both cloud and traditional.

Getting permission to dig is not easy. This is the "big deal" sale. As we have said repeatedly throughout this book, the big deal may not result in any revenue, but without it, you will never even get the chance to plant your tree. Your sales force clearly owns this job. Using the techniques we outlined in Chapter Eight, your sales rep will need to lead a team that can provoke interest, prove value, and negotiate the initial consumption contract. They have to qualify the customer to ensure that the growing conditions are good ones. They must work with the customer's line-of-business and IT leaders to decide what the initial usage profile is going to look like. Which apps? How many users? What levels of usage? That is all part of the initial contract. That is the responsibility of sales . . . no debate.

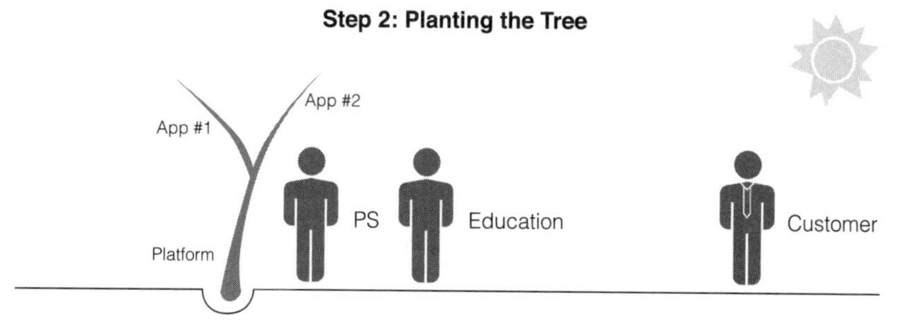

FIGURE 9.4 Plant the Tree

Great. We got permission to dig. So here comes the team of professionals who will get this baby in the ground. Hopefully the initial sales contract included all the right services to get the product implemented correctly. This would include the needed configuration, integration, education, and consulting to get the customer operational according to the initial apps and user agreements. We know customers don't buy everything we sell on the first pass. They pick our core platform and one or two of the key applications we offer through it. So that is what we are diligently working on implementing.

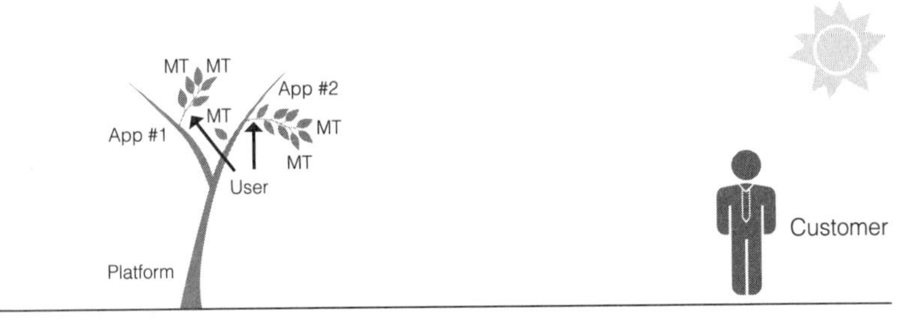

FIGURE 9.5 The Old Definition of "Done"

Okay, it's in the ground. Mission accomplished. The platform is in, the apps are functioning, and the end users are starting to consume the new capabilities. We likely even have a few micro-transactions starting to flow from the early adopters. The senior leaders feel like they got what they paid us for, a working and highly available technology solution. It's time for all of the members of our team to move on to the next customer and repeat the process.

That's what we are going to do, right? Isn't getting the customer to achieve "go live" the end of our sales and professional services role? Certainly in the old CapEx purchase model the answer would be yes. Aside from any tech support they might have needed, we were done. The problem is, the tree is hardly pretty; it is not yet vibrant, and it is likely a long way from being well loved.

The party on the hook for making it pretty, vibrant, and well loved in the old product playbook model—the one who took on the risk of getting value from the platform—was the customer. In the old model, we had our money, and we performed our contractual obligations to deliver and implement the product. Based on our old CapEx business model, it was time for us to move on to the next tree planting.

But this isn't the old model we are talking about. This is the cloud, where consumption, not availability, is the thing of beauty.

In the age of Consumption Economics, we don't want trees that look like the one shown in *Figure 9.5*. It's alive, but it isn't producing many pretty, blooming MTs. What we actually want is to grow trees that look more like the one in *Figure 9.6*.

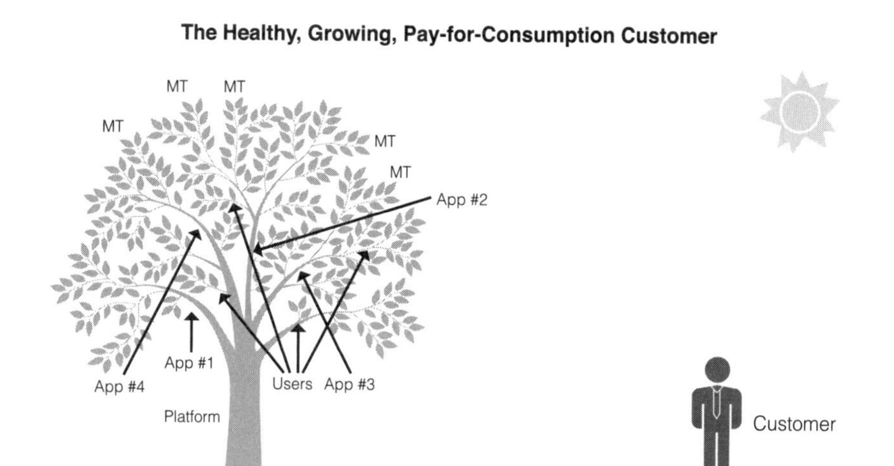

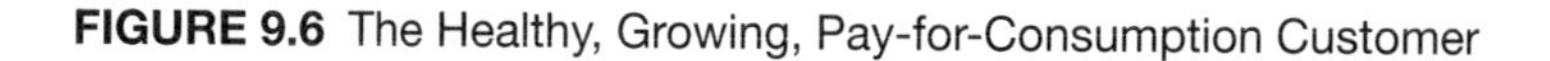
FIGURE 9.6 The Healthy, Growing, Pay-for-Consumption Customer

This is what we want. It is also what the customer wants. We don't want just a few good users, but depending on the product we sell, we want tens, hundreds, or thousands of good users. And we want to sell the customer not just the two apps specified in the original platform sales agreement, but maybe four or five apps that would enable the customer to really maximize their realized business value. In addition, we want to provide all kinds of user add-ons, like upgrading to higher-level feature sets, adding data or content services, and buying end-user specific apps from our ecosystem partners.

There is one other thing we want. Volume. Lots of transactions, lots of files being created and moved about. Tons of processing time and stuff to securely store. Lots of MTs for the whole MT pyramid of cloud (and on-premise) suppliers that make up the total IT stack for this customer.

So who owns that job at our company? We know that the sales team owns selling the trunk (the base product platform), and professional services and education services own planting it in the ground. But who owns growing the tree once it's planted? Who owns driving more limbs (apps), branches (users), and leaves (MTs)?

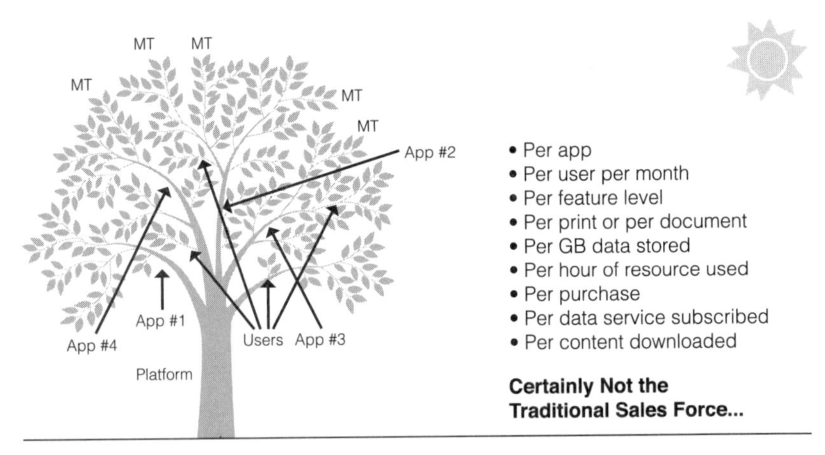

FIGURE 9.7 Who Has a Cost Structure That Can Sell at These Prices?

The question is a hard one to answer for a very specific reason. The revenue tied to growing an individual limb or branch or leaf on our tree is not very big. As we have said often, these might have no revenue associated with them, or they might have teensy, tiny micro-transaction-level revenues associated with them.

Oh, and one other thing—our quarterly financial performance is going to depend on these limbs, branches, and leaves. Our CEO wants to know who owns the forecast and performance for this revenue stream. The challenge is that we don't know of a single enterprise or consumer tech company that has a good answer today. The one thing we can all agree on, however, is that this is not the best use of our main sales force. Their cost structure is prohibitive, the skill sets required to sell these MTs don't align well, and the interest simply isn't there to pursue this type of high-volume, low-revenue-per-transaction business. Maybe our current inside sales teams could take on the selling of the apps or the upgrading of user licenses in bulk sizes, but that still falls far short of the total number of transactions necessary to deliver the revenue we need and the adoption the customers need.

Our sales "hunters" want to hunt big game, not grow little leaves. Their job is to get permission to dig from the IT and business execs at our customer accounts. That in itself is going to be mission-critical. Let them do what they do well.

So our CEO still has a problem: no gas pedal to accelerate the Consumption Economics engine. Heck, there may not even be an engine under the hood. So how do we solve that problem? Well, let's take a look at our existing assets and figure out what options we may have.

If you look at most tech companies' organizational entities, you see departments that are built to optimize around a specific kind of interaction. At one end, we have organizations like sales where we emphasize having the proper human knowledge and skills that can be used in critical one-to-one interactions with important customer decision makers and influencers. At the other end of the scale, we have organizations like product development that are tasked to build a model and technology infrastructure (our product) that are going to be highly scalable and fully optimized to be one-to-many. All other departments lie somewhere in between these two extremes.

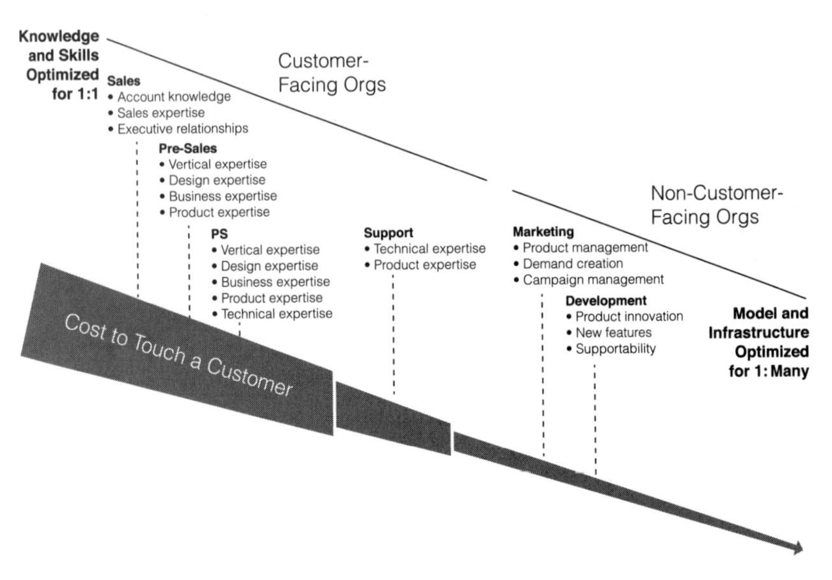

FIGURE 9.8 Current Expense per Customer Interaction

All seven departments have different cost structures and different costs per customer interaction. They are purpose-built to have an expense per customer interaction that is as low as possible but still have the necessary capability to execute the task or goal that they are chartered to perform. So while they all try to have the lowest possible costs, they also make the investments needed to ensure adequate capability. If any of these organizations cannot perform their tasks well, the whole customer process could fail. We need salespeople who can get orders done. We need professional services people who can handle the design, configuration, and installation challenges. We need support people who can diagnose and resolve complex problems. And we need marketing and development people who can create great products and generate demand for them.

Tech companies have spent decades striking the best possible balance of capabilities and cost structures for each of these organizations that, collectively, own customer success. So what's the problem? The problem is that we built all these structures around the old product playbook. They are designed to optimize our overall profitability in a business model that assumes up-front customer purchases and long-term maintenance annuities.

These organizations are simply not optimized to drive MTs. They are not well suited to selling a large number of low-priced apps, single-user licenses, small add-ons, or feature-level upgrades. Most of all, they are not suited to drive volume and product adoption. This is problem number two, and it is a big one.

How do we tackle this problem? Let's start by looking at *Figure 9.8* again. Below the individual department boxes you will see a pyramid labeled *Cost to Touch a Customer.* You also will notice that the pyramid has two major break points. These two break points represent the place in the overall customer interaction model where there are quantum reductions in the cost per touch.

The first is between professional services and support. That's because of three simple facts. The first is that most sales, pre-sales, and professional services interactions take place at the customer's site, while most support interactions are done from remote support centers. The second fact is that the sales, pre-sales, and professional services people make a lot more money than the support guys do. That was fair because the customer interaction tasks owned by sales and professional services were both complex and critical to getting the revenue. The tech support and field service folks had important jobs, but their individual customer interactions did not have the same importance. The third difference is the number of technologies and tools at their disposal. Customer service and support techs are armed with an incredible array of productivity tools that enable high levels of efficiency. By contrast, the sales and professional services folks have to rely mainly on their experience and knowledge—wetware, not software. This limits the number of tasks they can complete per day and their rate of productivity improvement over time.

Put these facts together, and you have the first big break point in our customer interaction economics.

The second big break point is between support and marketing. This one is really simple. Marketing and product development don't have much interaction with individual customers. Do they present themselves to the customer? Absolutely. But they do it indirectly. Development does it through the product, and marketing does it through their campaigns and one-off customer briefings or advisory board meetings. These departments are the ultimate one-to-many organizations. So this second big break point in our customer interaction economics is easy to understand.

So if we are going to build the capability to proactively grow huge and profitable customer trees, where are we going to start?

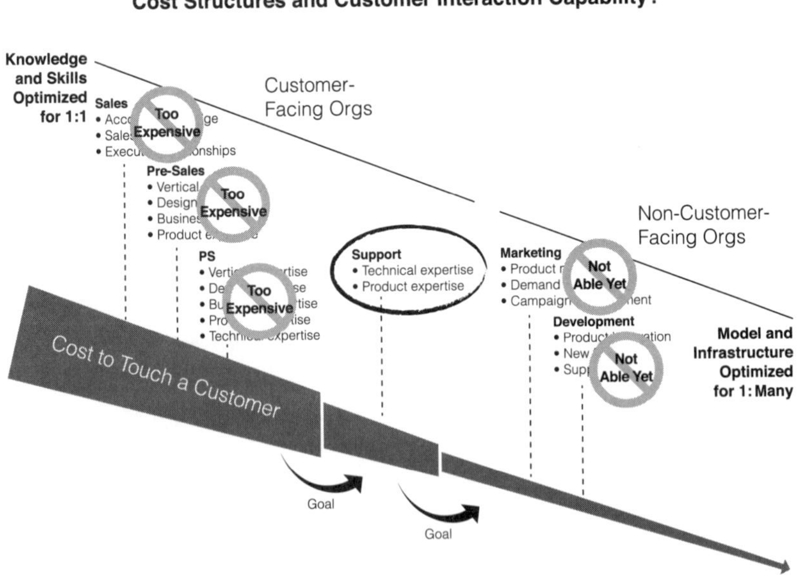

FIGURE 9.9 Who Has the Best Combination of Cost Structures and Customer Interaction Capability?

Having sufficient capability at the lowest possible cost is once again the right consideration as we build our new Consumption Economics gas pedal. We need to have adequate capability to really monitor and drive the customer's consumption of product value, but we have to do it at a profit.

So where are we going to start? Where is the best existing organizational asset for this purpose? Which department could best act as the basis—the foundation, if you will—for the new capability we have to build? Unfortunately, every organization in the chain has some level of deficiency when it comes to pushing the Consumption Economics gas pedal. The reality is a bit of a "Goldilocks" problem. We have some organizations that are too expensive, which includes our current sales, pre-sales, and professional service organizations with their expensive people and less-evolved productivity tools. And we have other organizations that are not really capable yet, who don't have the right customer-facing processes, which includes development and marketing. To be clear, it is absolutely our goal to push as much of the responsibility for driving consumption into the product and its embedded Consumption Innovation and In-Product Upsell layers as possible. The ultimate nirvana would be to not need people for the process, but to do it all through marketing and development. We just aren't anywhere near that point. We do need people to work with customers. But which people?

Which organization is currently the best positioned to be tweaked and optimized to drive Consumption Economics? We think the diamond in the rough is customer service and support. They are the perfect foundation on top of which to build our consumption gas pedal. They have the lowest cost labor model of any of our customer-facing organizations. They already have call centers all over the world. They are phenomenal users of productivity technology and tools. They already own the customer self-service websites. They are the best-formalized knowledge managers in the company. And when it's absolutely necessary to go on-site, they still maintain the lowest interaction cost in the company.

Customer support is an under-leveraged corporate asset. Most tech companies have spent billions of dollars to build it and optimize it. The result is a sophisticated and capable customer-facing organization. And you know what we do today with this amazing asset? We pump low-value content through it.

Customer service and customer support are there to fix broken things—broken products and broken customers. They keep the products up and running but rarely add any additional economic value to the customer. No one asked them to. It's like we built the space shuttle and then used it as a crop duster.

It is time to repurpose this asset. It is time to take this already-paid-for, state-of-the-art customer interaction machine and give it a new and much more valuable purpose. Customer support—in fact, all of your services organizations, including professional services, education services, managed services, and outsourcing services—need new missions and new responsibilities. They need to own growing the trees. They are not able to do it yet, but what they are missing is fixable.

Before we talk about how to move forward, we need to look at the slightly absurd services reality we have lived in during the product playbook days.

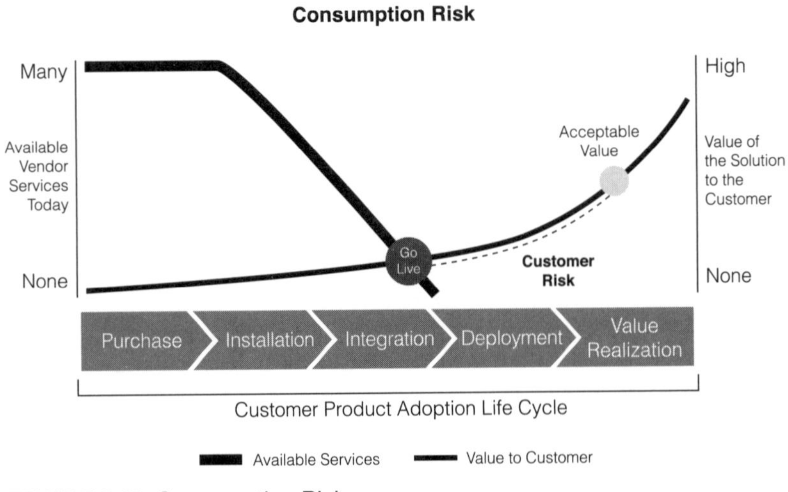

FIGURE 9.10 Consumption Risk

In the past, we have spent a lot of time and money trying to build our direct and indirect models for getting customers through the early days of the product adoption life cycle. We invested in building a sales force that could assist them in their purchase process. We built our professional service teams or developed a channel of partners who could help customers get their products installed and operational. In short, we had plenty of services available to get the customers to go live with our products, get their money in our bank account, and get their revenue recognized.

The irony of it was that at the point of go-live, at the very point where the customer was going to actually start accruing value from the use of the product, we were done. We pulled our professional service teams out and sent them on to the next implementation. Now the customer was largely left alone to get value from their investment . . . or not. They were the general contractor who had to actually deliver the result, not us. We had our revenue and our money. We had them locked into maintenance. Our economics were optimized. But the customer was at risk. They might get to the desired and acceptable level of value, or they might fall short.

And please don't allow anyone to kid you that owning the responsibility for customer value realization is the job of our current customer support organizations. Our current CFO-driven message to the support team is to minimize engagement with the customer, find ways to keep the product up and available using remote technology, and be preventive, predictive, and automated. If they do have to interact with the customer using people, we tell them to get in and get out. We talk about things like call deflection, minimizing talk time, preventing truck rolls. As much as the support staff might like to, they aren't trying to get in there and help drive value; they are trying to minimize costs and maximize service margins. They do their absolute best to delight customers, but they usually have to stop when it begins to cost extra time or money. Even very well-intentioned CEOs and CFOs are forced to adopt this stance as their product businesses get near the Margin Wall.

So the bottom line is that just as the customer enters into the critical adoption phases of the life cycle, our service offers dry up.

As we've said many times, the result is the Consumption Gap that almost every customer we know suffers from. The economics just didn't encourage the tech companies to get in there and deliver the full result. The legacy is today's shelfware, unused features, category commoditization, and unrealized value. In the old economics, none of these were near-term financial issues for the tech company. Under Consumption Economics, they will all come home to roost.

Well, it is time to change the engines on the planes.

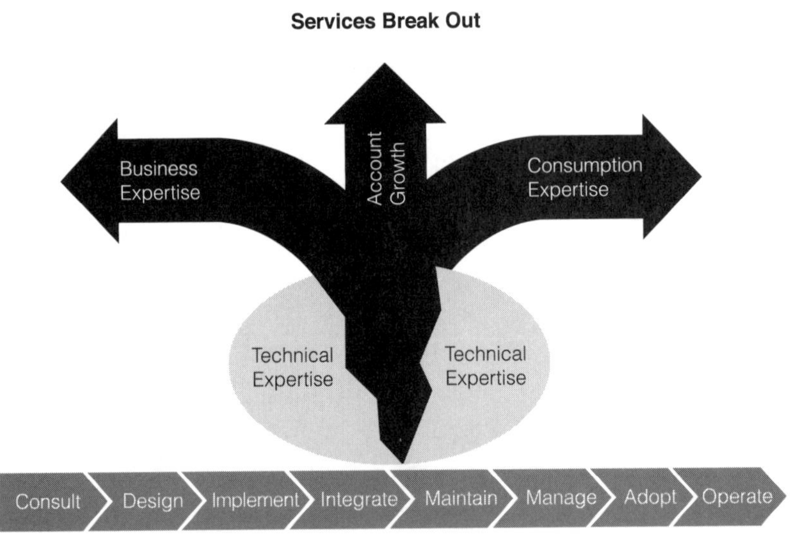

FIGURE 9.11 Services Break Out!

Services of all types—professional services, education services, customer support, field services, managed services, outsourcing services—need to move beyond technical expertise. They need to move beyond installation and maintenance. They need to move beyond being cost centers. They need to move beyond being considered simply "product attachments" by Wall Street. They need to break out of their shells in order to protect and defend their current golden revenues, and to answer the company's call for someone to own "growing the trees" in our cloud customers.

How are we going to do this?

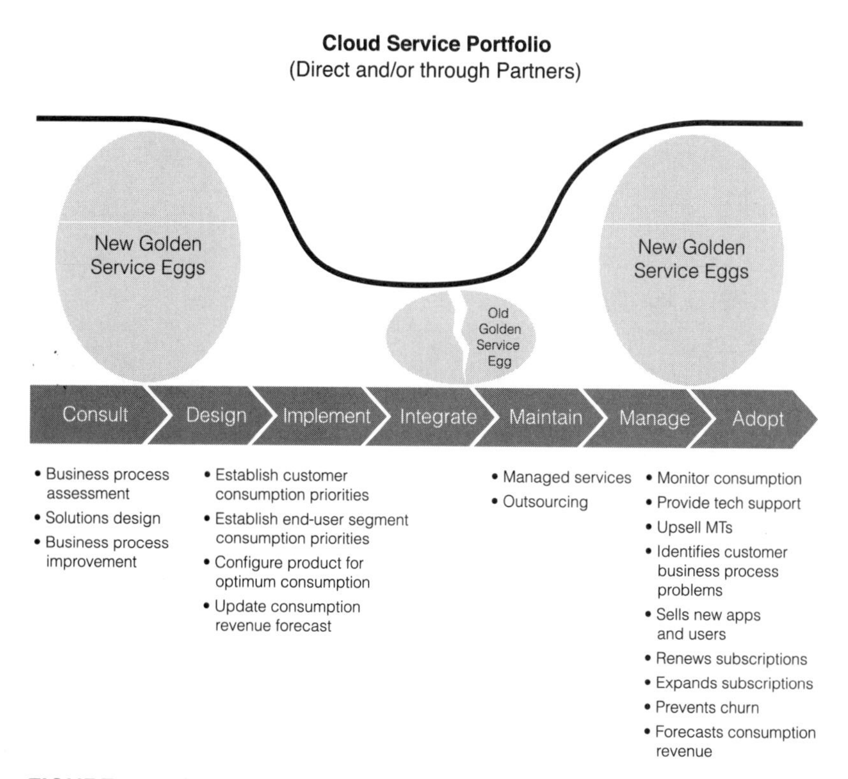

FIGURE 9.12 Cloud Service Portfolio (Direct and/or through Partners)

We are going to move services from the center of the customer life cycle to both of the edges. We are going to repurpose the asset toward new skills and new offers. We are going to take on consumption optimization and the realization of customer value as our main jobs. And services organizations are going to end up owning the customers and owning "the number."

In short, we are going to remedy the slight absurdity in the customer adoption life-cycle model.

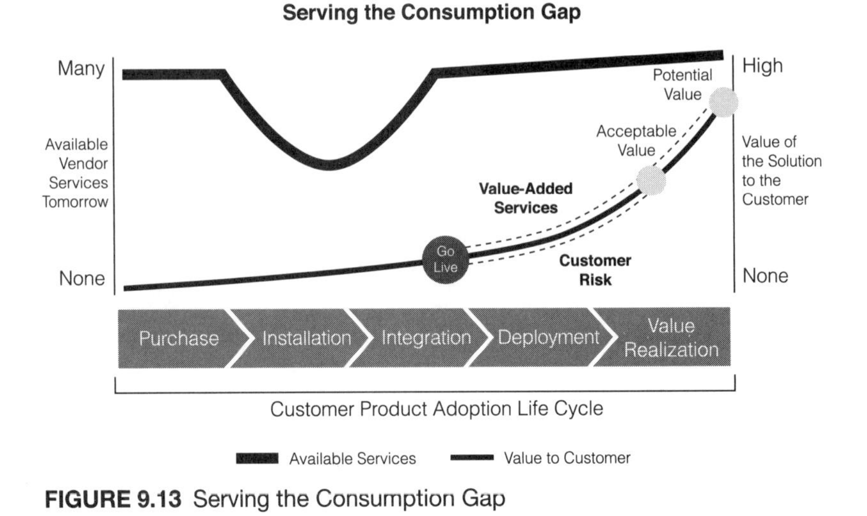

FIGURE 9.13 Serving the Consumption Gap

The goal of Consumption Services is not just to remedy the risk that customers took to get the product to a minimally acceptable level of value. The goal is to take them all the way; to help every customer realize the maximum business value that our whole product and services portfolio can provide. We need to align our service assets with the true requirements that our customers have—to help business buyers get business outcomes and help end users get personal productivity and enjoyment outcomes. In achieving their goals, we also will achieve our own.

The goal for the future would be to build a new, integrated capability. One that we will call the Account Services Organization (ASO).

The Account Services Organization is the melding together of existing service capabilities and new inside sales capabilities into a single, value-adding team. It will be built on top of the current customer support and field service infrastructure. That means the locations, the tools, the processes, and the labor base, all of which will need some retooling.

Introducing the Account Services Organization

Convergence
- Built on the customer support and field service infrastructure
- Includes the customer support and field service functions
- Extends infrastructure under the professional services and education functions

VAS
- Re-missions the asset to driving customer consumption of product
- Connected to Consumption Marketing and Consumption Development organizations

- Adds the ability to sell both in-bound and out-bound
- Owns the product subscription renewal

Mission: Monitor and Optimize Customer Consumption of Product Value

FIGURE 9.14 The Account Services Organization

The ASO, or whatever you choose to call it, is best thought of as "customer service plus." It will still include all existing customer support and field service job functions. It will still fix broken products and broken customers. But the goal will be to eliminate the need for those functions as fast as possible. Those tasks won't add value to anyone; they will just prevent disaster. Let's continue our drive for improved product quality, peer- and self-support options, usability improvements, and proactive, preventive maintenance solutions. As we do all these things and drive our break/fix caseloads down, we need to retool the assets to get into the Consumption Services game. We need to immediately start engaging in service tasks that add value to our customers' businesses.

The new ASO will have close ties to other service organizations like our professional services and customer education departments. They, too, should be brought on top of the same infrastructure as the ASO through a process of services convergence, which we will cover shortly. This is especially true as it relates to tools and technologies. The content that powers all services functions—professional services, support services, education services, and managed services—is becoming more integrated. It is needed not just in a single time or place. What we mean is that education content might be needed in year two during a customer support interaction, not just during initial product implementation. Similarly, a small amount of integration assistance may be needed in year three. We need to be able to give the customer whatever services content they need at any time and through any channel.

All of these service functions are shifting in exactly the same direction: away from technical and toward business. It just makes sense that they would not only begin to improve their efficiencies by leveraging the same tools, but that they would also begin to jointly build the next generation of business-oriented consumption knowledge.

The ASO not only will be tightly tied to the professional services and education organizations, it will be closely integrated with Consumption Marketing. As we pointed out earlier in the chapter, once the ASO has figured out how to isolate, duplicate, and process a Consumption Service task, it will need to push that task to marketing and development for inclusion in our in-product marketing campaigns. Our ASO also needs tight ties to development to get our people the heck out of the low-value, break/fix transactions through improved product quality and better built-in service tools. This allows us to take those tech support resources and redeploy them to growing more revenue by producing new limbs, branches, and leaves.

What Does Account Services Do?

- Monitors the consumption of customers
- Provides reactive, in-bound (IB) tech support (remote, online, on-site) and upsells MTs
- Does preventive tech support
- Provides proactive, out-bound (OB) consumption services (remote, online, on-site) (aka VAS) and upsells MTs
- Identifies consumption business process problems and forwards to CPS for customer action
- Sells new apps and users via IB and OB telesales
- Renews existing product subscriptions
- Expands existing product subscriptions during renewal
- Scans and prevents churn of MTs, users, apps
- Scans for platform churn and sends to Sales for customer action
- Forecasts all consumption revenue

FIGURE 9.15 What Does Account Services Do?

Figure 9.15 provides a general summary of the key responsibilities the ASO could assume.

This brings us to the two major new construction projects that the ASO will face.

Let's be honest. One thing that the current customer service and support asset is not great at is *selling*. This is a function that really needs to be built from scratch. Even if the sales team has set the stage properly for driving lots of micro-transactions through the customer's purchasing process, someone still has to get the end users to hit the "buy" button.

You may already be asking yourself, "Why is the ASO built on the customer services infrastructure and not the inside sales infrastructure?" There are several reasons. First, this is selling inside the act of supporting, not vice versa. Second, the service/support infrastructure can handle the deep, complex content associated with optimizing product use that basic CRM just can't. Their knowledge management tools, diagnostics, remote access, etc., are simply more advanced and far better suited to the kind of work we are going to be doing. The customer service organization can accommodate global accounts and follow the sun. They have better ties to product development. The list goes on and on.

But what they can't do is the selling motion. Is the best way to add this to merge the current inside sales team into the ASO? Maybe. Or maybe you let it stay in sales and keep it acting in support of the field sales teams. It depends on what they do today and what you need them to do in the Consumption Selling model you will build.

The key break point between what account development activities reside in sales vs. the ASO may be in selling additional applications onto the core platform. These are the limbs of the trees. As an example, how does salesforce.com get an existing CRM customer to take on Chatter or their ServiceCloud apps across the enterprise? That deal may be large enough for a dedicated inside sales organization to worry about—maybe even turning it over to the field sales force if the opportunity is big enough. But selling the branches and the leaves? No way. Adding just one more user, getting a user to move up to the next functionality level, adding a data subscription? That has to be done at a cost per interaction that only an ASO backed up by Consumption Marketing can achieve. Even selling additional applications to SMB customers might be better aligned to this new ASO cost structure than traditional inside sales.

Melding sales capacity into existing customer service organization capabilities is more complicated than just hiring a few people. There is a huge cultural shift that must take place. Just like we had to transform thc product sales force to work effectively in the new normal, we are faced with a similar challenge in the tech support and field service organizations. Among the current workforce, we will have some who view sales as someone else's job. Others will shirk from interacting with the business users. More will feel out of place having to talk about anything other than technology.

The second scary part of re-missioning customer support to own growing the tree is around the lack of business expertise in the current workforce. This is another major shift. It's interesting that we build vertical market products, have vertical marketing sales forces, and even have vertical professional services people. But support is organized by severity level with product escalation paths. That is not the right path forward. Broken products in the cloud are not support's problem; they are development's problem. We need whole new layers of skills in the Account Services Organization.

Again, just like in the sales-force transformation, some will come by these new sales and consultative abilities quite naturally while others can be re-skilled. And once again, there will need to be a pruning exercise. We can't emphasize enough that the over-arching thought that this organization needs to have is a deep belief in our product's ability to help the customer. The more apps they can get the customer to purchase, the more users who use them, and the more add-ons they consume, the more business value the customer will capture. The job of the ASO is to give the gift of that business value outcome to every customer. This is not slippery selling; it is more like a doctor who wants his patient to be as healthy and vital as he can be. If you are in customer service today and reading this makes you feel uncomfortable, we would invite you to consider that maybe it's not the role that is making you queasy; it's your true belief in the value of your current product. If you truly believe that your products are valuable, you should not have any ethical problem promoting their consumption. If you don't believe that, you shouldn't walk away from the role, you should walk away from the company.

The ASO is the answer to the gaping hole in most companies' Consumption Models. Don't worry—we can feel your skepticism . . . corporate confidence may be low when considering the idea that something built on top of customer support could one day own the customer. But the hard reality is that no one currently performs this role in the old product playbook model. Either we have to build something from scratch, or we have to adapt something we own. Think about your alternatives, and we believe you will arrive at the same conclusion we have: Customer support is the best foundation to build from.

Looking at the Broader Consumption Services Engagement Model

Once sales sells the platform to the customer, the Consumption Services teams should take over. First, the Consumption Professional Services Team should prep the product and prep the customer for success. Then the Account Services Organization should take on the ownership of the customer. It will support them, monitor them, and grow more limbs, branches, and leaves. At the end of the day, they will own the number for that customer.

They will forecast that customer's total consumption and be responsible for delivering those numbers. Along with their partners in sales, they will be responsible for the size of every tree.

We could write a whole book about all the things that must be done to build and manage Consumption Services in the age of Consumption Economics. For now, however, we just want to go on record as saying that this is the way forward. It is what salesforce.com means by calling their support plans "Success Plans." Their website describes them like this:

> *When you're using Salesforce, you want to get the most value. We offer a unique combination of support, training, and administration expertise to promote user adoption, increase productivity, and ensure high ROI.*

That is also why you shouldn't call any of these new organizations technical support or even customer service. The name of the organization at salesforce.com that supports, renews, and grows customers is called "Customers For Life." That is its formal name. Whether it is a full-blown Consumption Services function yet is not the key point. The point is that they saw what we see: a gaping and mission-critical hole in the world of Consumption Economics that did not exist in the old CapEx-based product playbook model. We don't believe any company in the world is doing all that they could be doing to optimize consumption. Executing Consumption Development, Consumption Marketing, Consumption Selling, and Consumption Services, including the full-blown ASO, is a tall order. It is a journey, not a project. But we can say that the SaaS and IaaS companies we work with are starting to coalesce around some realities of the new normal. It is a different business model with different stress points, different economics, and different success factors than the old models we all grew up executing.

Executing a Consumption Services strategy does not simply entail creating a successful ASO. We also need to reconsider the role of professional services, education services, and managed services. Let's start with professional services.

Every professional services organization we know has some version of these three issues in play today:

1. How do we automate and offshore (or at least off-site) our low-end technical project work?
2. How can we create more reusable and re-saleable IP?
3. How do we re-skill our workforce to step up to business consulting where required?

These issues are not limited to a company's cloud offers; they are being driven by the internal push for higher margins and external evolution in customer demand. When TSIA thinks forward to professional services in the age of the Consumption Model, we focus on these things:

Introducing the Consumption Professional Services (CPS) Organization

- Business process assessment
- Solutions design
- Business process improvement
- Establish customer consumption priorities
- Establish end-user segment comsumption priorities
- Configure product for optimum consumption
- Update consumption revenue forecast
- Normal PS functions

Skills Inventory
- Vertical industry expertise
- Business expertise
- Design expertise
- Product expertise
- Technical expertise
- Consumption expertise

FIGURE 9.16 The Consumption Professional Services (CPS) Organization

At the risk of over-simplifying the change in objective from the old professional services model to the new Consumption Professional Services model, we would say that in the old one, professional services had the job of technically enabling the product for use. In the CPS model, it is getting the product *and the customer* ready for consumption. What do we mean by this distinction?

For years, professional services has been about installing, integrating, designing, configuring, customizing . . . all the technical things needed to make the base product operational. The goal has been creating a functioning implementation of a product according to a given customer's deployment spec.

Professional services organizations are now realizing that their true value is not just technical expertise; it is business expertise. That is what the IBMs and the HPs are learning with their advanced consulting teams. Sure there is plenty of complexity still out there, and that can be what your service business is focused on. But what many companies see is that business expertise and consulting are the next big wave of service value. TSIA has observed several embedded professional services organizations that are explicitly reducing product implementation and integration activities by migrating them to partners. Why? So the internal professional services resources can focus on activities that help customers align business objectives with technology capabilities. Services such as these:

- Consulting services. These services leverage the tech company's knowledge and capability in their area of expertise. The services involve diagnosing customer problems and opportunities, and solving them through innovative technology and business process improvements. Think IBM Smarter Planet projects.
- Design services. These services take defined business objectives and translate them into IT systems. They rely on system architects who are not only deep in a single product, but who also can understand the whole ecosystem of technologies required to achieve the objective. They are willing to learn the customer's infrastructure, their dependencies, and their constraints. They deliver the solutions specification that best fits the objectives. Think Capgemini projects. Or think about

what happens inside Best Buy when a consumer comes in to buy a home theater system.

- Operate services. These are services where the tech company not only supplies the technology, but also assumes responsibilities that could range from managing and maintaining it to actually providing the employees that use it in a production environment. There is a huge range of offers in this exciting and high-growth area of services including new types of managed services and outsourcing services.

The CPS team of tomorrow will still own all the technical implementation services that remain in the cloud world (which should be less than in the on-premise model), but they also will inherit some important new tasks inside our overall Consumption Model.

In CPS, we need to focus on getting the customer optimized for consumption using a spec that is partially ours and partially theirs. We have tons of knowledge of business process optimization around the end-customer functions that our products perform. They want us to help them with that. We know what can be done with our products in ways that any one of our customers simply cannot envision. We get to see an "*n* of many" while they are an "*n* of one." They want us to show them those insights. We know what questions to ask, and we are going to ask them.

Remember from Chapter Seven and Chapter Eight, where we talked about building Consumption Roadmaps that were a reconciliation of two inputs: our own Best Consumption Practices recommendations and the Desired Consumption Practices that emanate from an individual customer and its end users, as seen in *Figure 9.17*?

CPS owns the creation of that second model. We need to work with our business and IT buyers to help them identify their preferred end-user and organizational consumption priorities. We also need to poll key end-user segments to see what features they

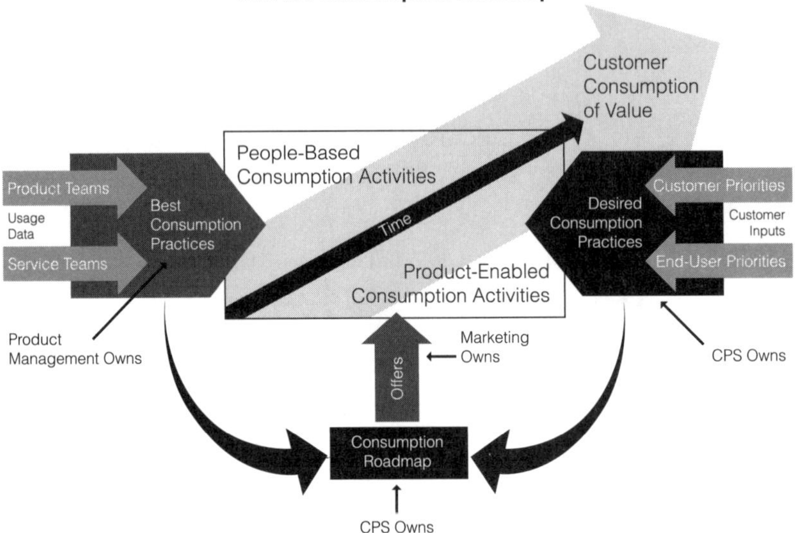

FIGURE 9.17 CPS Owns the Desired Consumption Practices and the Consumption Roadmap

prioritize. Together these inputs must result in that customer's Desired Consumption Practices.

Next, CPS must reconcile both of those models into a specific Consumption Roadmap for that one customer. There will definitely be conflicts. Some corporate customers could want to emphasize functionality that does not lead to the stickiness that our product managers aimed for in their model. Maybe our Best Consumption Practices don't apply in this customer's case.

What really matters is that all the best thinking is on the table for this one customer's deployment of our solution—the best thinking of our internal experts, the customers, and the cumulative experiences of all the other customers who have gone before them. The moment where this all comes together to the benefit of a single customer is magical and rarely, if ever, happens today. It has never happened this systematically, at this scale, and this early in the customer's product usage life cycle. This is the true power of the Consumption Model. It will lead to faster realization of core product value, higher levels of advanced product value, lower internal help-desk costs, lower vendor management costs . . . all things that make their internal business case for our product sizzle like never before.

Once that critical roadmap exercise is complete, CPS needs to direct the product on how to carry it out. As we mentioned, development will have created a whole next generation of product configuration functionality focused on interface command and control. We need to configure the product according to our Consumption Roadmap; to make it appear in the right way to the right end users. We need to document the user segments whose consumption activities we want to track. We need to set up the first few waves of new feature and MT offers according to the plan. In effect, we need to educate the product on how to present itself to end users in the way that all our hard work and analysis have suggested will achieve the best consumption outcomes. CPS will be the people best suited to use the new Consumption Innovations built in by development to do the initial product interface configuration and establish all kinds of product rules and goals according to the Consumption Roadmap. This step should be part of their initial implementation engagement.

Once all the technical and consumption implementation steps are complete, the customer will go live. Hopefully sales has included enough CPS hours into the agreement for them to actually shepherd the customer and the Consumption Model through its first weeks. We might need to change some of our initial command-and-control decisions. We might see positive end-user adoption patterns that we want to encourage even more. There are all kinds of tweaks we could and should make to optimize consumption (and their economics) during that first month.

Then it is time to pass that customer on to the Account Services Organization for care and feeding. As we said, it is their job to grow big trees.

Before we review the end-to-end process, we also want to say a quick word about the future of customer education. Customer education for most tech companies is a line of business. We sell classroom and online training about our products for a fee. In the world of the Consumption Model, education will often morph into a value proposition delivered through the product and our service teams rather than remain a standalone offer.

Yes, there may still be some amount of basic training that gets sold up front. But let's face it, "just in case" training—that is, training every end user and system manager on everything they will ever need to know about our product—has never worked. Students are lucky to retain 20 percent of what they learn in a course. Like so many other things we have discussed in this book, it was the best we knew how to do. We trained them on the whole enchilada up front because we knew we were unlikely to ever get them back into the education setting again. But it just doesn't work. It is very inefficient.

Once again, our Consumption Model allows us to correct much of that inefficiency by making the product start off simple and unfold in real time as the end users are ready to consume more advanced functionality. We will train them in real time, in small chunks, delivered through the product and our Account Services Organization. We will shift from just-in-case to just-in-time. Since we now own responsibility for driving consumption, we will make sure that every feature and offer, every service and support transaction, every moment of idle screen time educates a user. Education becomes a ubiquitous flow, not a five-day classroom ordeal. As an example, and yet another interesting intersection between consumer technologies and the enterprise, training is moving from all-day sessions to three-minute iPhone apps. Need to learn about a new VMware capability? Go download the app and watch it over a cup of coffee. Then go do it.

We realize that all this may have organizational and financial implications for tech companies with large education business units. But once again, e-commerce companies lead the way. No one ever went to classroom training to learn Amazon,

Walmart.com, or iTunes. You will know you have executed your Consumption Model perfectly when your customers don't need your classroom or remote training either.

Okay, so what might the whole end-to-end consumption cycle look like? How will our company plant big, beautiful trees at our customer's site without any of our hardware or software being there? We think it looks like this:

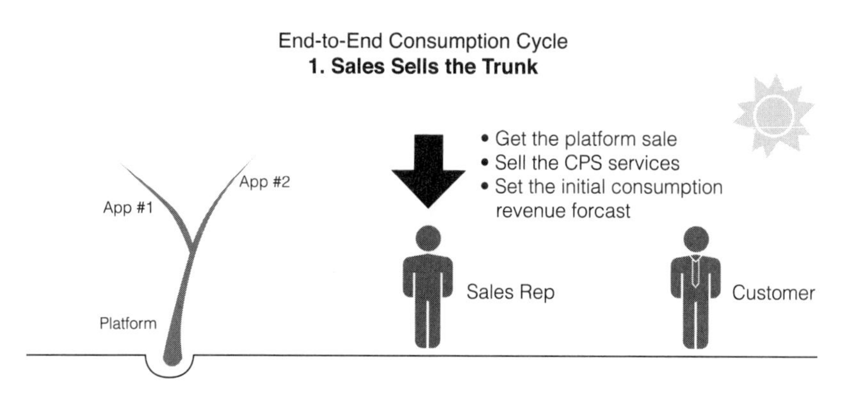

FIGURE 9.18 The End-to-End Consumption Cycle: 1. Sales Sells the Trunk

FIGURE 9.19 The End-to-End Consumption Cycle: 2. Consumption Professional Services (CPS) Preps the Customer

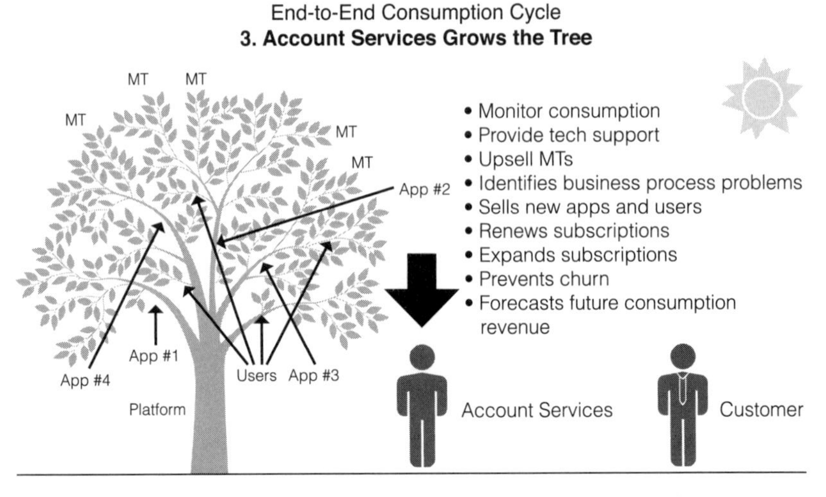

FIGURE 9.20 The End-to-End Consumption Cycle: 3. Account Services Grows the Tree

Once we have executed this cycle we will have a beautiful tree with lots of apps, users, and MTs. The ASO will be able to arm sales with all kinds of great consumption and outcome data for this one customer. Sales can then leverage our success into selling our second platform or major new application, and repeat the cycle.

As we said in the beginning of this chapter, the transformations coming to sales, marketing, development, even CPS, will be challenging. But they are transformations. The closest thing we will face to new construction will be building the Account Services Organization. The foundation is in the ground, but the footprint is not quite right. The structure on top has some great pieces in it but needs a new, more modern design. And at least one whole new wing needs to be added.

An important concept from earlier in this chapter needs to be reiterated here. Our goal is not to thwart the process of automating human customer interactions just because we have built an effective Consumption Services organization. We should remain conscious of the overriding objective: to embed as much of the process as possible into the product's Consumption Innovation and In-Product Upsell engines.

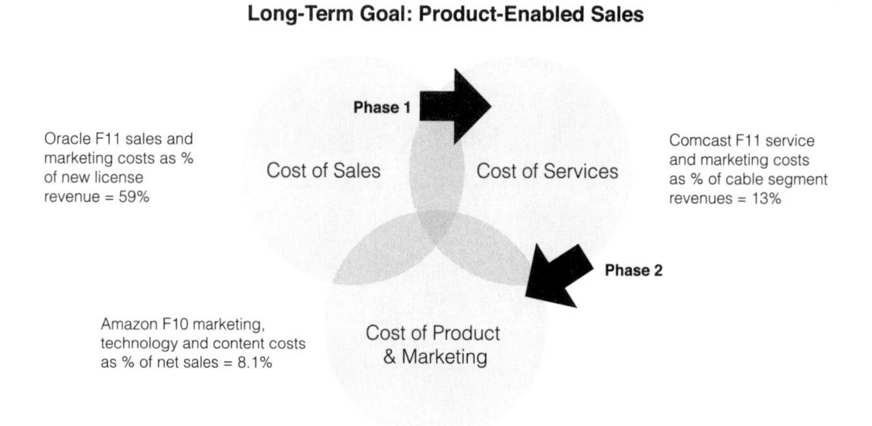

FIGURE 9.21 Long-Term Goal: Product-Enabled Sales

The ASO, in particular, is a chance to move *some* of the economics of complex technical sales off the traditional sales-force cost model and down to the lower break point we referred to back in *Figure 9.8.* That is progress, and that is probably the best we can do for now. What we are selling is not simple. It is not selling books or office supplies. Difficult decisions remain for customers to consider, options to compare, situations to fit. Customers, and especially end users, are going to need some help from experts.

Look at the cost of sales, services, and marketing figures we reference in *Figure 9.21.* We already discussed Oracle's high sales and marketing costs as a percentage of new license sales. We chose Comcast as our example of a company who has moved their sales cost off to their customer service platform. Beyond that, Comcast has done a great job of trying to package their rapidly increasing complexity. They started off with a very simple choice: basic cable television or a premium package that included HBO. Remember those days? Now there are separate cable, voice, and Internet offers. For each of these lines, there are advanced, premium, pay-per-view, or similar complexities. It is getting darned complicated. Yet none of us has ever had a cable sales rep show up at our front door. They use marketing to create demand and customer service to sell. Their customer service and marketing costs are 13 percent. That's a great cost-of-sales range to target for the ASO.

Why not go straight for the Amazon economics, with their 8.1 percent cost, which includes all of their marketing, technology, and content costs? We would argue that it really doesn't even work perfectly for them, and they are one of the best at executing their model. They still have abandoned shopping carts due to complexity. Many times these are complexities that a human could have overcome. We are just not ready to fully automate technology sales even in e-commerce, much less in enterprise tech. We will need humans in the process, and the ASO represents the best chance at doing that at an affordable cost structure.

Another key question that we will wrestle with is how this new function fits into the organization chart.

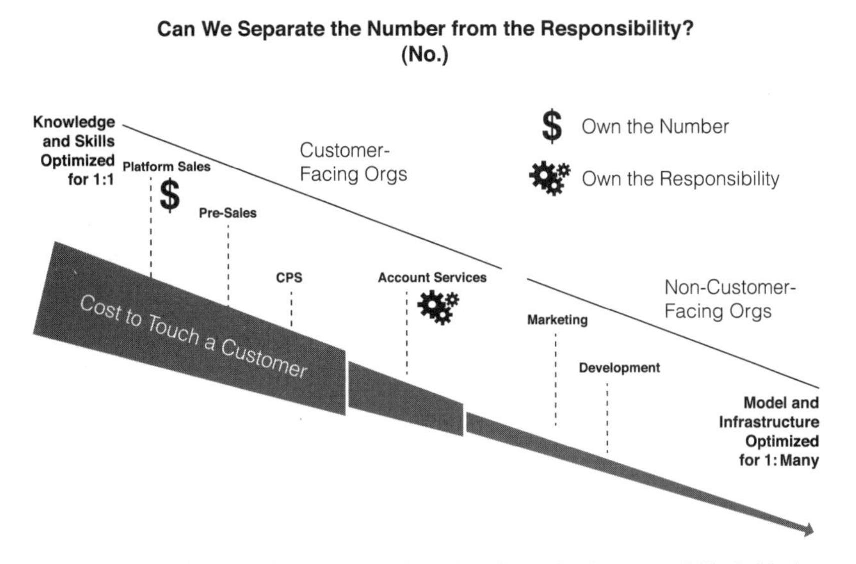

FIGURE 9.22 Can We Separate the Number from the Responsibility? (No.)

Because the customer service organization has never been thought of as being sophisticated at the sales motion, many companies will be nervous about trusting them with critical revenue streams like SaaS contract renewals and upgrades, new app sales, add-on sales (MTs), or even adding end users. To those companies, the idea of customer support (or a derivative of it) actually owning total account revenue must seem insane. They see the function as purely reactive, with no genuine account development skills. They still call them "techs."

That is a fair criticism, and one that the services executive community must respond to. But the fact is that we have to stop trying to divorce the selling motion from the service motion. In the cloud and other Consumption Economics models, the service motion is the optimum selling motion. Service calls won't be about down systems, they will be about how to do something. This is a sales opportunity, and the person who provides the answer will be that customer's trusted advisor. At this moment, the sales labor costs are optimized. The prospect is on the line and in the moment. This is the time to sell.

Some companies will exhibit a vote of no confidence by leaving the revenue number with the sales executives even as they shift the execution to the ASO. They will correctly realize that putting the ASO directly under sales will change its nature, compromise its altruism, and force it over the line. They are smart enough to realize that it would break the sacred trusted advisor covenant with customers. Despite knowing this, they will use metrics to effectively have it be owned by sales. We think that won't work. Over time, these organizations will realize that separating the delivery from the ownership of the number results in underperformance.

We believe that the service executive community can and will stand up to the challenge. It makes sense for them to own the number for existing customers. They are in the best position to forecast it and the best position to deliver it. But these execs need to sign up for the program. If not, then they force a choice for the company, and it is not a choice they will be happy about.

So how do we make the transformation from the separate service silos that we have today to a united sales and service function capable of owning the existing customer base?

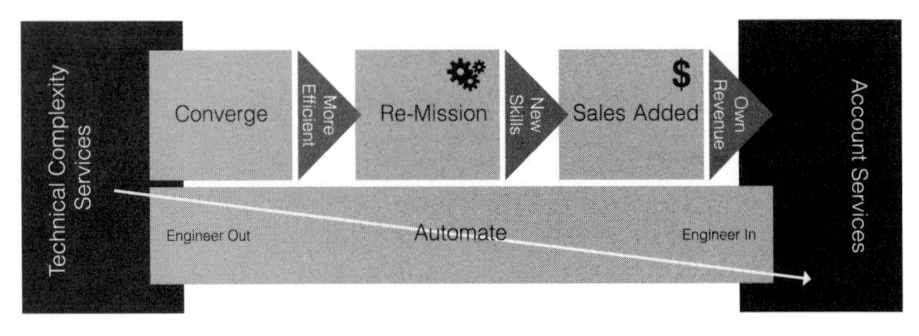

FIGURE 9.23 Transition to Account Services

Think of it along two tracks. The big-picture journey is to repurpose the service business asset from solving problems created by technical complexity to focusing on consumption and account development.

The first track is automation. We need to automate virtually all of the existing technical services. We need to engineer *out* the need for human service intervention using a multitude of strategies like improved product quality and usability, automated configuration and integration tools, preventive maintenance, remote support, etc. We know the drill; we are just under-investing in it. Then we need to aggressively engineer *in* automated Consumption Model and sales functionality. All the things we discussed in Chapter Six.

The second track is the more delicate one because it involves lots and lots of people. We have divided this track into three stages:

1. Services convergence.
2. Re-mission the assets.
3. Add a sophisticated sales function.

We talked a lot about services convergence in *Complexity Avalanche*. At a simple level, the goal of convergence is to seek efficiencies and economies of scale by intelligently and selectively combining all the separate service organizational silos we see today in most tech companies. This should free up cash to reinvest in the rest of the services transformation.

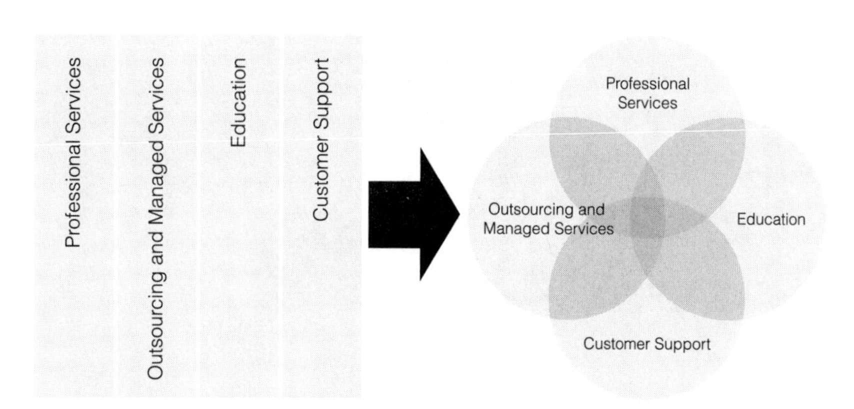

FIGURE 9.24 Services Convergence: From Silos to Sharing

The idea is that there are many redundancies and inefficiencies in the silo service model. This is theoretically easy to imagine. But the real power behind service convergence is the ability to have the service functions not just share infrastructure, but share knowledge and put the right resource on the right task without the limits of organizational structure. Why isn't there more professional services content on the support self-service website? Why can't a field service rep perform a low-level professional services task while they are at a customer's site to fix a box? Why can't a professional services person who is not currently assigned to a project ("on the bench," as we like to call it) be temporarily tasked to build content for customer education or edit new submissions to the knowledgebase? Why can't we extend the case and field management technologies that support has worked so hard to optimize into the professional services, managed services, and outsourcing services businesses?

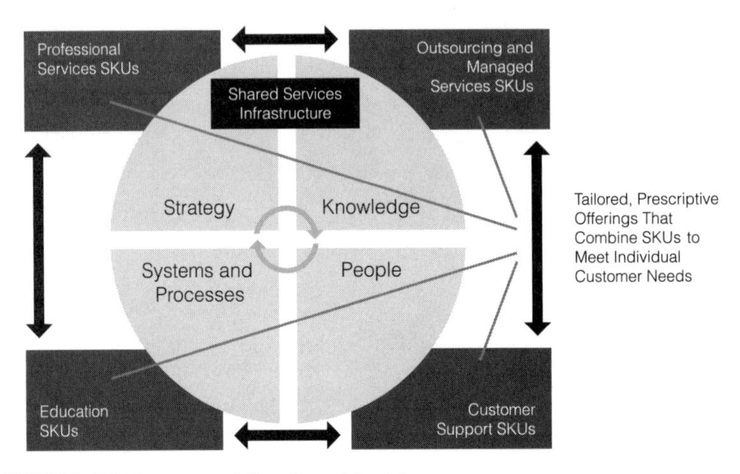

FIGURE 9.25 Converged Services Model

These may make sense to a greater or lesser degree from one company to another. We would maintain that service convergence could contribute to improved margins and value creation at almost all companies. But like anything else, it is a political battle. The center circle in *Figure 9.25* is the key. We need a single service strategy, not four separate ones. We need to combine systems and processes wherever possible. Knowledge created by one silo should be available to all the others. And the SKUs, meaning the individual services tasks contained in each silo, should be easily combined into a single proposal for a customer. Do you think for a minute that a customer who needs some combination of professional services, education services, and tech support to improve their usage wants to talk to three service salespeople, have three separate diagnoses, and then get three separate proposals? No.

The people part will be the hardest. We all know the internal politics and cultures that make it tough. Once again the uniting thought needs to be that our interests and our customers' interests are aligned in the world of Consumption Economics. Anyone anywhere who can drive more consumption is a hero and needs to be recognized as such. This begs for a much more flexible workforce management model that fully leverages every services employee.

You can read more about services convergence in *Complexity Avalanche*, but we do have to say that we see too many companies who are attempting it today making the same mistake. They are often misinterpreting the "duality of peers" intention of convergence toward the low-cost provider. Here is a critical chart from that book.

Extending the Core Competencies of the Existing Silos

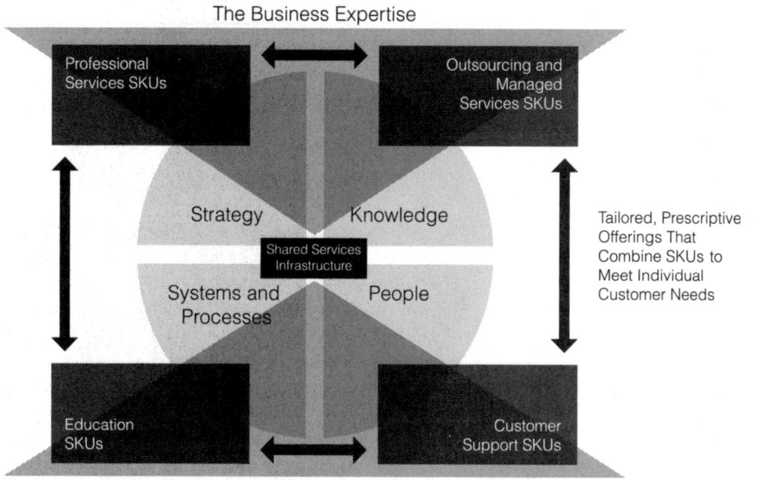

FIGURE 9.26 Extending the Core Competencies of the Existing Silos

The idea is not that customer support takes over other functions simply because its costs are lower. That's like saying the factory manager gets to pick what products the company manufactures. No, the idea is to let the manufacturing expertise of support and education build and manage the factory aspect of all the service business, but let the professional service, outsourcing service, and managed service teams determine what they do and how profitable they are. As we move away from technical complexity services and toward business services, we believe we will move more toward the "product manager guiding the factory" model that we see in manufacturing. They are two different skill sets, and they both have value. The idea that one is dominant over the other is wrong.

As an example, we have seen companies who are trying to move professional service sales deal desks under support just because they are call centers. Hold on! What does support know about professional service sales deal desks? That would be like saying the field sales force should report to IT because they use laptops and CRM. Sure, move the deal desk to the call centers and let them use their tools, but the content, value, and results of their work—including the people themselves—still have to be owned by professional services.

There might be some service functions that are combined or reorganized due to convergence. Certainly it should come together at the top of the company, reporting to the CEO. But we can't over-converge just because everyone is using the same factory.

Think about a new normal service model that looks like the one shown in *Figure 9.27*.

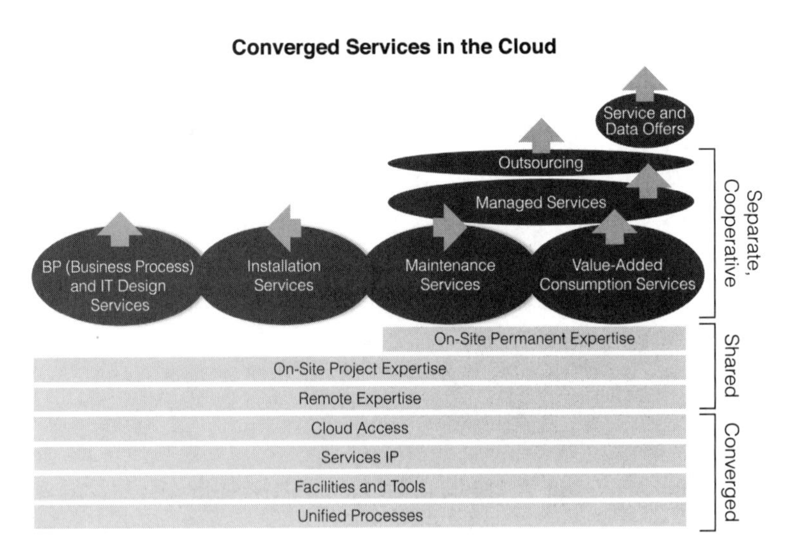

FIGURE 9.27 Converged Services in the Cloud

All of the service functions are working on top of unified processes and tools, which can be accessed by customers and employees in the cloud. They are all contributing to and refining the same IP. They are sharing expertise, both remote and on-site. People can be shared across offers and can contribute to the advancement of the converged infrastructure. The offers and the delivery people inside them are still managed by separate line-of-business execs, but they are cooperative and collaborative.

Some current product companies are on board with many of these concepts. HP's software business just launched a truly converged service offering that brings together what used to be separate maintenance and professional services functions, then adds multivendor software support to make a truly unique, next-generation services offer.

The point is that the services model is changing quickly. The winners will be those who act decisively and thoughtfully. Those who cling to the old product playbook too long will perish.

There is one final conversation about services that even we three authors are still in debate over. We call it "BoS vs. CoS."

A few pages ago, we pointed to some potential danger in the redirection of our gross margin source from old-model service annuity revenue streams like maintenance to product subscription "________ as a service" (XaaS) revenue streams due to their historically different price elasticity. This is a warning for the executive who says, "No big deal, because I simply have swapped this profitable maintenance annuity revenue stream for an equally profitable annuity revenue stream called product as a service. I am just replacing one Golden Egg with another." But there may be a problem with this logic. A potentially big problem—at least if we use history as our guide.

Remember our chart in Chapter Three illustrating the Margin Wall? In it we have an arrow that represents the typical trend for technology cost/price over time. It is pointed down. As we emphasized, that trend, which is well known in the hardware world, is now coming to software. And for all the reasons we have mentioned in this book, it will happen to the cloud. It will not only happen, but it will happen very quickly. It already is happening today. As those prices decline in sector after sector, the per-customer profit margins could—especially without a robust Consumption Model in place—decline right along with them. After all, you will still have fixed costs in procuring and servicing these cloud customers. So if you think that you can just swap one high-priced, high-margin revenue stream for another, we are raising the possibility that it isn't that simple. Flipping the Golden Egg customer from the existing service annuity stream to the cloud revenue stream could mean lower margins for the industry. If we aren't careful, we could lose the Golden Egg and have nothing to replace it with. There are already some industry estimates that the cloud could shrink total IT spending by as much as half.

Here is why we are thinking about this: In the old model, we separated the service revenue stream from the product revenue stream. That turned out to be a brilliant move . . . maybe the best single move ever made by the industry. Why? Because it sheltered the margins of a gigantic revenue stream called services from the highly competitive, discount-oriented product business. Even when the margins in certain product categories went down to near zero, we survived and prospered because we could rely on a profitable service business—one that we could monopolize, one that we could count on year after year through good economic cycles and bad ones. As we said in Chapter Three, it may have muted the true effect of commoditization in many market sectors.

Well, here we are at the dawn of the new era called "cloud." And here we are facing a fork in the road. Are we going to bundle the services that customers will need into a single subscription price and hope to prove the investment return on ASO just through increased product subscription revenue alone? That effectively makes service a cost of sales, which would be just one more cost to get the customer's product business. Or will we be smart enough to continue to separate services as a standalone revenue stream?

BoS or CoS?

Business of Services
- Monetize service opportunities with customers
- Break out service revenues and costs above the line
- Services contribute to the top and bottom line of the company
- Services will be positioned as a source of growth and profitability

Services as Cost of Sales
- Services are included in product subscription
- Services are COGs
- Product subscriptions are positioned as the primary source of growth and profit for the company

FIGURE 9.28 BoS or CoS?

This is a critical question for tech companies that are already in or are moving toward the cloud or any other utility/subscription/service-based product pricing model. We wish we could give you the definitive answer.

Let's examine your options: One strategy is to continue charging separately for services. Let's call that Business of Services, or BoS. The second strategy is to bundle all the basic services into the product subscription price. Let's call that Services as Cost of Sales, or CoS. So BoS or CoS, what is your strategy going to be?

Let's start with the advantages of BoS. The most important advantage of the BoS strategy is that it retains the tech company's ability to protect a large portion of its total revenues from the vagaries of competitive product pricing commoditization. In this model, you will package, price, and sell your services separately from the product subscription price. You will account for the services revenue and profits as a related but adjunct P&L. You will invest in your service capability because it is a source of profitable revenue growth. And you will leverage your services capability as a source of competitive differentiation.

Sounds great! Why wouldn't you do that? Well, the CoS model also has its advantages. Mainly, it's simple for the customer to understand. Charging a flat fee per unit of consumption (per user, per transaction, per gigabyte, etc.) and having all the necessary services be part of that flat fee is exactly what customers say they want. It is also what most web-based companies in the consumer world do today. Early SaaS companies have gone this route, including basic maintenance in the product subscription price. They then charge separately for advanced services. The best practice is not yet known.

Further adding to this need for simplicity is the confusing XaaS nomenclature we have settled on. The word "services" got hijacked by somebody—salesforce.com, we presume.

You see, services were traditionally recognized as being separate from the product. Just like the way that people taking action dominates the definition of service in the dictionary, people taking action was understood by tech customers to be a separate thought (and purchase) from the product. But now we are blurring the two in potentially unproductive ways.

If we end up going the CoS route on basic product services, everyone could lose. Shareholders could lose because "product as a service" commoditization would take down the whole company's margins, not just the product revenue's margin. Customers would lose because when margins head down, tech execs will do what they always have done: cut service costs. Would they cut the sales force? No. Would they cut their R&D spending? No. They would cut the service budgets. And that would cut consumption service availability and quality.

And get this new slant: The company itself will lose. Why? Because services is going to be repurposed to drive micro-transaction volumes in order to sell more product. If we are not careful to preserve that capability, we run the risk of cutting off a key driver of top-line growth. In order to make the capital markets happy with our margin percentages, we risk starting a cause-and-effect cycle that starts as a margin problem and ends in stagnant growth. In addition, in the very near term, services will become the customer's trusted advisor. Product salespeople around the industry will cringe at the thought, but it is absolutely going to happen. Customers are losing interest in speeds, feeds, and complexity. "Get me the result" is the new RFP. And that is a services-led discussion.

In the cloud we may need to retain and defend the idea of separate services. Tech services that bring experts and their tools to the aid of customers still have value and need to be invested in. Remember that the cloud is based on successfully closing the Consumption Gap. We must drive adoption and utilization. We must engage and take responsibility for proactively driving customer success in ways that customers cannot do on their own. We need to do that in order to improve our own growth and profitability. We must also do it to overcome the Consumption Gap and enable the best possible business outcomes for our customers. Getting customers to pay us for that just might still make sense.

One final thought about Consumption Services. At the spring 2010 TSIA Technology Services World conference, we interviewed Geoffrey Moore on stage. He made a key point that every tech executive should insert into his or her thinking. He said, simply, "Services is an asset that can increase the return on all other assets." In 2010 at the dawning of this next generation of tech business models, he probably didn't know how right he was!

Services *is* an asset that provides a greater return to all the company's other assets than almost any investment short of the product itself. Services can increase customer revenue, improve product margins, cut cost of sales, improve market share, increase market adoption, and play the role of trusted advisor to even your largest customers. We need to make the shift to value-added services both to defend and protect the golden service revenue stream for as long as possible and to fill the gaping hole in our cloud customer strategy. One move, a re-missioning of an existing asset, can solve both of these critical problems.

It is time for services to break out. Customers are asking for it. Shareholders will reward it. It is up to the capital markets and the C-suite to let it happen.

10 Customer Demand vs. Capital Markets: *How Fast Should You Transform?*

THE REACTIONS OF PEOPLE WE HAVE TALKED TO ABOUT THE concepts in this book fall into three categories. There are the *Believers*, who see Consumption Economics as an opportunity to better serve their market and disrupt slow-moving tech business models. They want to get there as fast as they can. Then there are the *Skeptics*, who reject the idea that the cloud is going to alter their corner of high-tech anytime soon. They are heavily addicted to the CapEx purchase model—it is how they make money. They are planning to talk about private clouds and hybrid models for as long as they can and hope they cash out before Consumption Economics really takes hold in their markets. Lastly there are the *Pragmatists*.

This chapter is for them.

Predicting the future is one thing, and we are pretty confident about what's coming. Calling the *timing* of when the new model will become the standard in a given market sector or geography is much more difficult. The Pragmatists agree with the core trends and their implications. They see that their cost structures were built to manage expensive complexity and just can't be supported at these lower price points. They see future growth coming less from the big customers they know and more from the little customers they don't know. They see the power of the concepts but are wondering how fast they can and should move toward the Consumption Model.

It's a delicate balance to manage—a high wire to walk without falling off. They have a product playbook business model today that they depend on for their current revenue and profits. On the one hand, they want to keep that going for as long as they can.

On the other hand, they see the vulnerabilities that could result from clinging to that model for too long. They are watching salesforce.com and Workday gain market share over SAP and Oracle. They have seen VMware seemingly come out of nowhere to capture a whole new market opportunity by being a stepping stone to the cloud. They wonder what Google will get into next. They see "consumer grade" devices being used where "enterprise class" was once the norm.

A senior exec at one of the world's largest tech companies commented to us: "We talk about all this new behavior we want and, as we get close to the end of the quarter, we tell all the sales reps to go out and sell boxes." That is the conundrum. There is often a balancing act between what our progressive customers want and what our short-term investors expect.

We think there are three key influencers that the Pragmatists must pay attention to. And the three may have very different perspectives.

- **Early-adopter customers.** Our product playbook business structures just creak and groan at the stresses that are being applied by the new wave of cloud-obsessed, early-adopter customers. If we don't rewrite the playbook into something more modern that's better aligned with their demands, our long-held customer relationships will simply break down and a new competitor will pick up the pieces.
- **Start-ups and innovative competitors.** Existing business models, organization structures, and the expectations of financial

analysts don't shackle aggressive start-ups. They are free to build new models from the ground up. They will organize around the Consumption Model from day one. These aggressive companies and their consumption-based cloud offers can be a scary threat to a current market leader who is still reliant on the product playbook. What's more, these start-ups can be acquired and brought into the portfolio of an existing competitor as an "end-around" play. The Pragmatists must keep a very close eye on competition because they can disrupt their natural path to the wall. They can take away the runway that you thought you had.

- **Capital market analysts.** How would the financial community react to a big public tech company that missed its quarterly revenue and profit target because customers who were forecasted to buy the old CapEx way instead shifted to the pay-as-you go OpEx way? How would analysts change their modeling, earnings estimates, and stock price guidance if this happened? What if it happened three quarters in a row? What would happen to Oracle's stock price if they announced a true 100 percent consumption-based SaaS pricing model and stopped charging maintenance fees? It's not that the capital markets have no model for valuing these models—heck, they are giving salesforce.com a price-to-earnings ratio that is ten times that of Oracle or SAP. But what would they do to you? The trick is in how you handle these shifts. How well can you demonstrate the total future profitability of your new cloud and managed service contracts vs. your old product models? How well can you predict the migration of your customers from the old to the new? What visibility and transparency can you provide to the financial markets? How well can you track progress against your new guidance?

We have to start the journey no matter what Wall Street thinks in the short term. We have to educate them on what our progressive customers are telling us they want and what the future competition will look like, and we have to smartly model out the financial transition. Then we have to get our transformation started ASAP. Think of it as a 12-step program to wean our company, the financial analysts, and even our shareholders off of the old model. Otherwise, we will run it until it fails. We will take our company and crash it into the Margin Wall.

How Long Is *Your* Runway to the Margin Wall?

The first assessment that tech executives need to make is how far away from the Margin Wall they are today and how long they have before they get there. Moving to the Consumption Model is going to take time. We need adequate runway if we are to fly over the wall without hitting it.

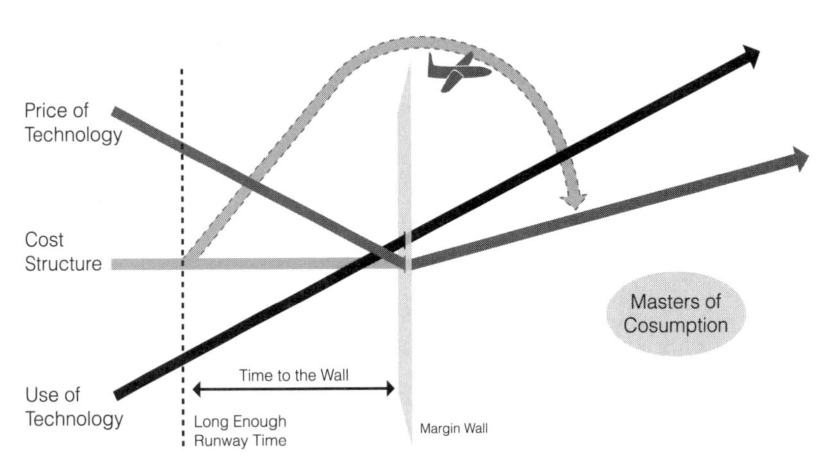

FIGURE 10.1 The Consumption Model Takes Time and Money to Implement

As we said, you could be staring at the wall today simply because the category you compete in has already commoditized. As we argued in Chapter Three, we think the Consumption Gap has actually commoditized many more categories than the players within them are willing to admit.

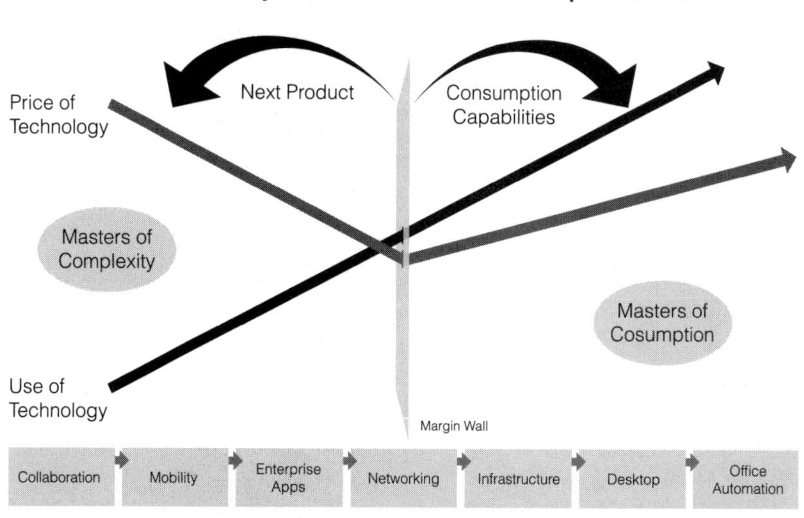

FIGURE 10.2 Market Maturity Drives Demand for a Consumption Model

If your company plays in categories that are already commodities, it is going to take some real belief that you can decommoditize your product from its peers by investing your cash into becoming a master of consumption. Apple did it. Xerox and GE Healthcare are working on it. If it can restore product margins to higher levels like it will for these companies, building your Consumption Model could easily pay for itself.

But let's be honest. It will take a few quarters of reduced EBITDA forecasts as you make the necessary investments. Unless you model out your transition in detail and set guidance conservatively, revenue and profit misses due to Wall Street's old-normal expectations are pretty much guaranteed.

IBM became the classic business-school case study to highlight the true breadth and depth of organizational transformation during its dramatic rebirth in the early 1990s. IBM hit its own version of the wall, and Lou Gerstner announced that they would transition the business model toward services. It was not pretty. One of the authors analyzed that transition several years ago. It took IBM well over four years to migrate to its new business model. During this time, the company took $8 billion in losses before stabilizing. Admittedly that was a different kind of transition. We would call IBM today the most new-school of the old-school players. They are using services to pull products, and that is progressive. On the other hand, they are positioning themselves as the ultimate masters of complexity, hardly the model we are advocating for most companies.

IBM has two unique angles. First is their ability to tackle the complexity of migrating the entire data center, not just one part of it, to the cloud. Second is their lab-based ability to create new software solutions to seemingly intractable problems. That is great for them. But for the rest of us who lack that depth and scale, we have to find our special place. Plus you have to ask yourself how long even IBM's runway is to the Margin Wall. They are the poster child for complexity and will go its way, good or bad. The real point is that these kinds of transformations take real time and real money. Yours will too.

Besides being in an already commoditized product category, the other reason you could approach the Margin Wall unnaturally fast is that your product category gets pulled into the cloud faster than you anticipate.

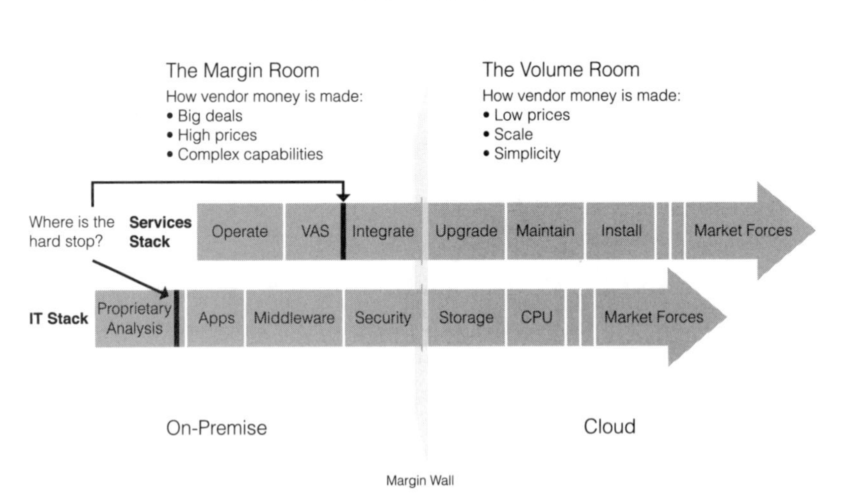

FIGURE 10.3 How Much Will Go into the Cloud?

You may not know exactly how long it will take before sizable numbers of new and existing customers embrace cloud or managed service versions of your product category. So depending on what business you are in, it could happen more quickly or more slowly.

We believe that within the next two cycles of technology refresh, most customers will move away from the old model. If your tech refresh cycle is typically three years, then our rule of thumb would say you have six years to be completely transformed. If your cycle is two years, then you have four. Given the scope of the changes we have been talking about across your development, sales, marketing, and services organizations, four years is not very long. And making it even more urgent is that new customers in emerging markets around the world may just decide to skip on-premise IT altogether. They might just go straight to the cloud.

Don't forget: Your transformation will be huge in both scale and complexity. Most companies we know are planning to start after they or their competitors launch the first compelling OpEx cloud or managed service offer in their category, because that's when new buyers will begin to sign up. We have to ask: Is that really when the race begins? Or is it even sooner?

How many good cloud products will never get off the ground because the product was ready for the new age but the rest of the company was stuck in the old one? It is one thing to have a version of your traditional product that customers can access off-premise, but does that make you a cloud or SaaS company? No. It makes you a traditional tech company with one toe in the air saying you are in the cloud. True cloud solutions need the different sales and service capabilities we have discussed in this book. Short of this, you risk looking like you are just a company who is worried about losing its position in the marketplace to some upstart SaaS company.

Worse, you might even conclude that your new cloud version doesn't work simply because your early adopters aren't happy or they aren't growing as fast as you thought they would. That is probably not the fault of your new cloud product; it is your lack of supporting activities. So can your Consumption Model transformation wait to start until your first cloud offer rolls out, or do at least the first-generation transformations need to be complete by that time?

A few months ago, a financial analyst was on CNBC talking about the cloud. He noted with a touch of irony that we (the tech industry) have spent the last 20 years putting our stuff into the IT department, and now we're going to spend the next five years taking it all back out. Is he right? Is it just five years from now when the size of corporate data centers and IT departments will be markedly smaller? That's not long, and it hangs together pretty well with our rule of thumb. Maybe the concern over security, the FUD (Fear, Uncertainty, Doubt) factor created by the on-premise IT leaders, or fear of change will slow it down a bit and give us more runway time.

Or maybe Intel and McAfee or someone else will solve the security problems. Maybe it will also prove to be cheaper to rip out an existing on-premise computing, storage, or software function and move it to the cloud than to keep it, pay maintenance on it, and operate it with internal staff. If those two things happen, then stand back! According to SYS-CON's *Cloud Computing Journal*, way back in 2008, EMC CEO Joe Tucci declared that 85 percent of data would be managed in what he called "big, safe information repositories in the Internet 'sky,' so to speak. We're [talking] cloud computing . . . "That was 85 percent, folks . . . not 8.5 percent.

On the other hand, maybe there are categories that will never move into the cloud. Maybe there are "hard stops" somewhere. That will be the last 15 percent.

No one can give you an exact date. That's why the first exercise in your Margin Wall homework is to calculate the length of runway you have in front of you. You have to study the tech refresh rate in your category. You have to look at how economically compelling the "________ as a service" value proposition is in your sector, especially with your aggressive start-up competitors. You also have to consider your brand positioning. Do customers expect you to be a leader on new trends? Are your early-adopter customers already giving you fair warning that you need to get going?

Maybe your company is not in mainstream IT. Maybe your company makes niche tech products like medical devices or industrial controls. If you are in a niche business, you should ask yourself how long it will be until the value-add in your product category moves from the hardware to the software side. Then how long until the software connects to the Internet? Then how long until the software value can come from the cloud and the hardware price commoditizes? That's the math assignment in your Margin Wall runway length calculations. And even if there is no imminent threat, maybe you see competitive differentiation or higher product margins that could come from a robust Consumption Model. Or maybe declining maintenance margins are keeping you up at night, and you see this as the right time to change your value proposition and save that revenue stream.

At TSIA, we call this shift to OpEx tech *The Natural*, just like the movie.[1] We think it is primed to happen, it wants to happen, it can happen, and it will happen. But it is up to you to calculate how much runway you have. That is your first important exercise.

How High Is the Margin Wall for You?

The second major assessment tech executives need to make is how difficult it will be to design and install your new Consumption

Model and how much it will cost to make the transformation. Think of this as the height of the Margin Wall that you need to fly over. The two subsidiary questions are, of course, how much will the transformations cost, and how long will they take to plan and execute?

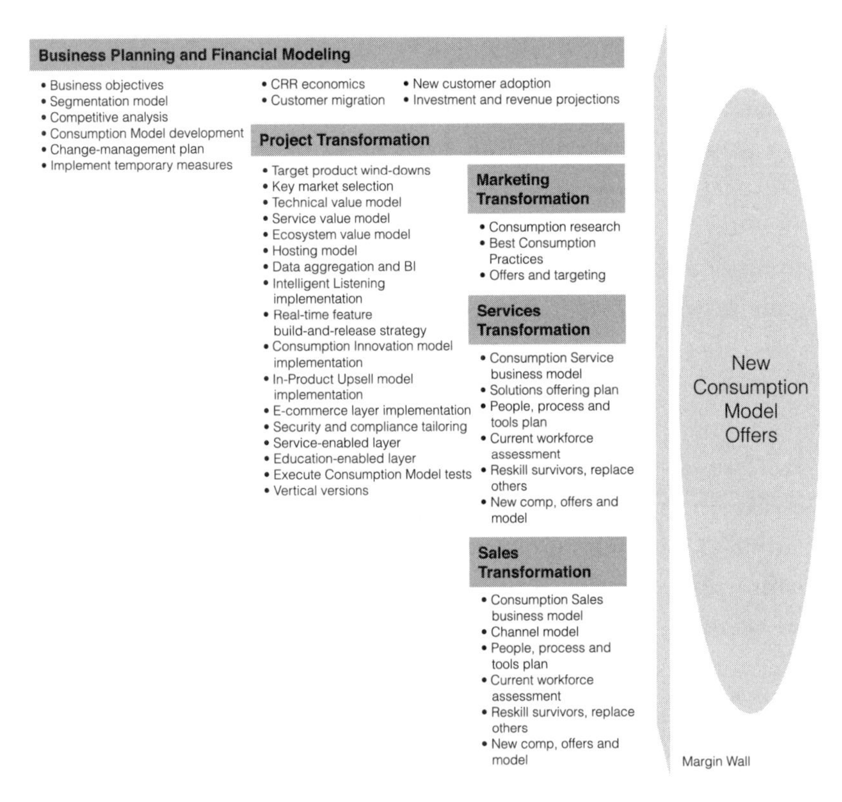

FIGURE 10.4 Sample Critical Path Investments to Fly Over the Wall

It is very hard for us to speak to the costs in a general way. They are so company-specific that any broad statements we could make would be fruitless. What size company you are and how you operate today could result in transformation costs that range from millions to billions. What we do have some experience with are the time frames.

Just developing the business plan, doing the financial modeling, and getting buy-in for the transformation could take six months to a year. The product changes could easily take two years. In the meantime, you can begin your sales, marketing, and service organizational transformations. Three years feels like a bare minimum to us. This sounds like an eternity in a cloud world. It is. But these are big ships to turn. That is why start-ups may have the advantage.

One thing is clear: Waiting too long to get started because of the fear over how Wall Street will respond can be catastrophic. Rather than beginning the journey, some companies will bet on their product playbook strategy to rescue them one too many times. They will hope and pray for a return to the world of complexity where they can live comfortably far away from the wall. With each passing day, this becomes more of a bet-the-company hope—not a plan.

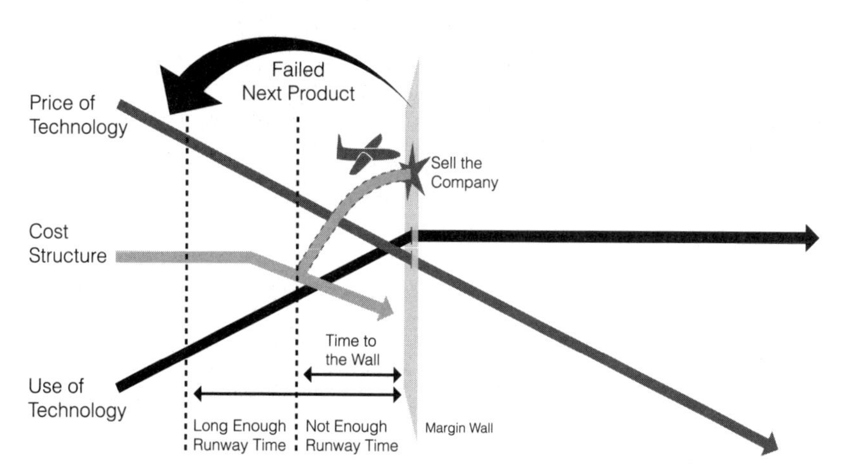

FIGURE 10.5 The Danger of Capital Markets Forcing the Too-Late Scenario

If that product playbook fails to accomplish the goal and you waste time and money on that effort, you may have taken away the runway and the resources you need to fly over the wall—just like what happened to DEC, SGI, and Sun. Let's be honest; what has the performance been for your last five "game changing" new products? How many have actually changed the game? Are you really willing to bet the company that the next time will be different? Company after company continues to spend billions per year in ineffective R&D, with very little new competitive advantage to show for it.

Your competition and your customers are moving to the other side. Apple is there. Salesforce.com is getting close. Amazon, Rackspace, NetSuite, Google, IBM, and HP are all pouring money into both their cloud infrastructure and their cloud business models. Dell is reinventing itself around it. You have to get there. You can't hope it will be on somebody else's watch. Consumption Economics is coming to a market near you. You need a plan yesterday and action tomorrow.

Buying Runway

You may find that your two Margin Wall homework assignments (determining how long you have vs. how long you need) leave you feeling uncomfortable. The math might be pointing to the fact that your runway length is not as long as your best-guess transformational time line. Or you may be looking at a rough transformation cost that you feel would be simply impossible for the capital markets to accept given the guidance that your company has already committed to.

In either case, you may need to establish some quick-return steps that can help mitigate the risk of your transformation. If you were a pilot, you might think of this as "buying runway" to get your Consumption Model off the ground.

Now, there is no silver bullet here. You will need a coordinated effort: two to three quick wins that can buy time for the

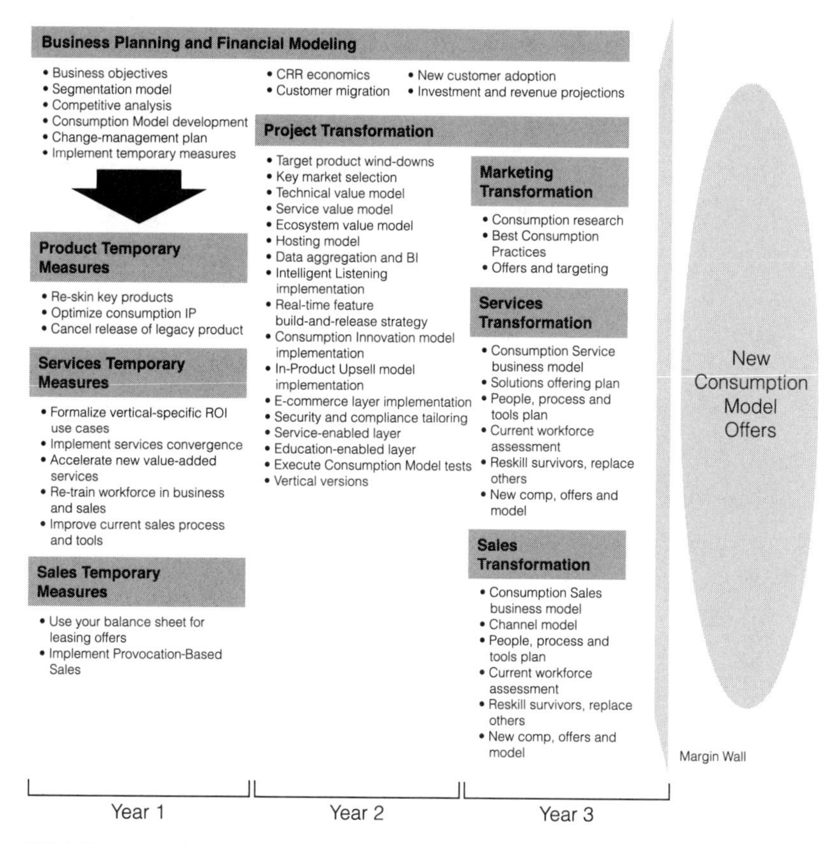

FIGURE 10.6 Sample Temporary Measures and Investments

longer transformations to take place. As examples, below are 10 near-term initiatives to consider.

The first three focus on your current products and offers:

1. Re-skin your highest revenue products to include alternate user interfaces for novice, advanced, and expert users. This may take three to four quarters, but it will provide end users with a way to build their consumption over time on your existing offers. It may even result in premium user levels that can

be sold at incremental fees. Think video games and how they reveal advanced features gradually instead of all at once.

2. Optimize your consumption IP. Your best customers and employees already know a lot about how to extract the maximum value from your solutions. Tap into this by actively developing customer, partner, and employee communities for your solutions. Appoint great technical and solutions leaders as moderators to take the average value of the content in your communities up instead of down over time. Reward content contributors for how many hits their content drives and what the average value rating is that other communities' members place on it. Use YouTube video or smartphone apps to promote your consumption IP to your customers and end users.
3. Consider canceling or substantially delaying the next release of one of your legacy products. This is a one-two punch. It frees up scarce engineering time to self-fund consumption-oriented investments, and it sends an important message across the company. It shows that you are making the turn from feature overshoot to a customer-centric development approach.

The next near-term initiatives may represent even more immediate payoffs. Focus your services and support organizations in new ways that directly increase the adoption of your existing offers:

4. A quick-win opportunity here is simply formalizing the highest ROI use cases of your offers in your biggest vertical markets. Where does your current offer have the demonstrated capability to bring a tenfold advantage to customers in a given industry? For each one, you need to get your services marketing or vertical marketing team to build CAR (Challenge, Action, Result) war stories that your sales and service teams can leverage to start speaking the business language of the customer. Once you have enough of these in a given vertical,

you can form a reference model for how a high-performance bank or retailer or government operates once they have fully adopted your solution.

Pilot the core service concepts included in *Complexity Avalanche*, which include the next four initiatives:

5. Consider the efficiency and productivity gains of services convergence. This also will lead to new service offers that reach across the old service silos to pull together innovative combinations that can better meet your most demanding customers' new-normal business needs.
6. Re-mission your field service, customer support, and professional services teams to actively engage with customers on capturing business value. Don't think that you can leave this to your ecosystem partners. They can't do it fast enough, and they lack the scale that you do. Focus these activities on your highest potential growth customers.
7. Invest in retraining your services workforce to increase their business, application, and sales savvy. Teach them ways to make every customer interaction not just a problem-solving exercise but a sales lead or customer consumption opportunity.
8. Improve your services sales processes and tools. We believe there are billions of dollars being left on the table because of weaknesses in basic service contract attach and renewal processes. This is true for both recurring service businesses, like maintenance, as well as project-based offers, like professional services. These are subjects that TSIA is becoming even more focused on simply because the best practices are so seldom employed effectively. These can range from sales-force design and joint selling models to basic customer database optimization.

Finally, there are at least two major quick wins to act on in your product sales organization:

9. The most immediate is to actively use your balance sheet to blunt the pay-over-time attraction of competitors' consumption-based offers. If customer CFOs want to buy over time instead of up front, let them. If you do not have the balance sheet to make that happen, seek out a partner who does. Again, this will buy time for your development team to design out the complexity that inhibits consumption of your offers and for your new Consumption Services to come online.
10. Last but certainly not least, get started on Provocation-Based Selling. The first step is to build compelling customer provocations around the high-ROI vertical reference models above. Maybe you just pick a single vertical like Sybase, Cisco, and ServiceSource have done. Maybe you can tackle three to five industries in parallel. Either way, you need to arm your sales teams with a new selling model and new ammunition to win the war for market share as consumption-based offers become the norm in your category.

All of these actions will help you "buy runway" to get your Consumption Model off the ground. You need a plan that balances these quick-win, current-fiscal-year initiatives with the longer-term, higher-risk investments required to transform your business. We are changing the engines in-flight. For the next few years, we will need to execute a "do-both model." We need to both run the current product playbook and layer-in a new Consumption Economics game plan. And we cannot afford to get either one wrong.

Consumption Economics

It's hard to describe how deep the roots of the product playbook model really are in our industry. It was the business model and set of activities that we all grew up with. It's where our current expertise lies. It's also deeply rooted in the capital markets. It is the tech model that those analysts know how to critique and how to value.

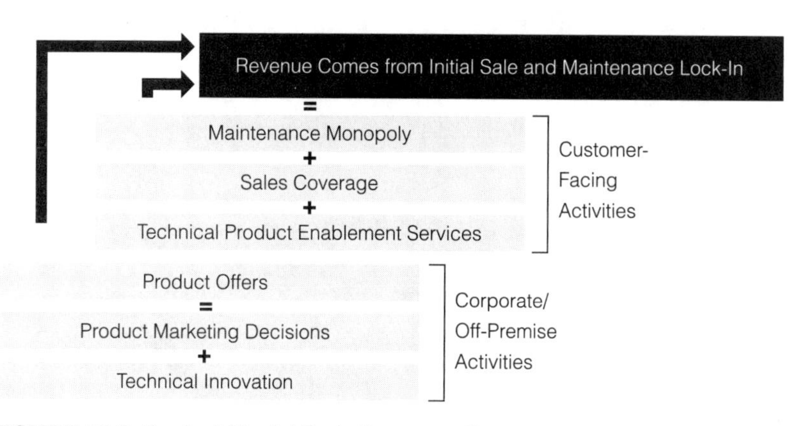

FIGURE 10.7 Product Model Tech Company Strategy

We understand tech business models that have technical innovation and sales coverage at their cores. They rely on high-margin product deals and maintenance or upgrade lock-in. Our managers know how to hire people who can build products, and those who can sell or service them. These key revenue streams were based on up-front commitments, not based on consumption. Ensuring that customers get value has truly been a nice-to-have, not a must-have.

But there are new rules emerging in tech.

At their core, companies that succeed in the world of Consumption Economics will look different. The Consumption Model must be built from the ground up to get customer results. As we have said repeatedly, we need to think about both product innovation and service innovation. These two thoughts must become an integrated, single thought.

Companies need to accept this new reality at their core. They must alter their DNA, and that is not an easy thing to do. They must build their strategy, their organizational capabilities, and their product designs to prioritize and differentiate around consumption. To accomplish this successfully, senior executives may need to re-think their offers, their go-to-market decisions, and even some parts of their workforce. But in the end, it will be a truly winning model, one that is modernized and ready for the next big wave in the industry's evolution.

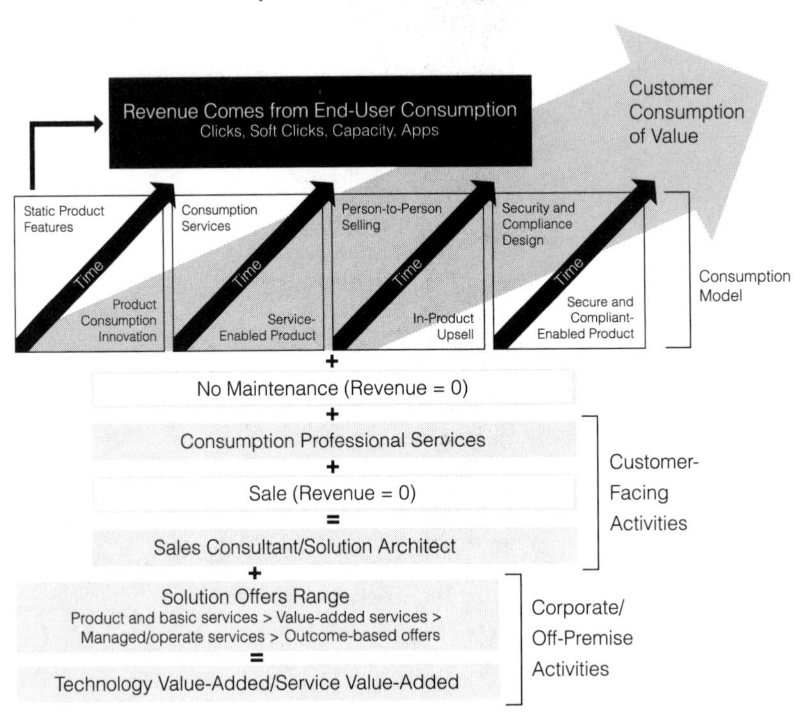

FIGURE 10.8 Consumption Model Tech Company Strategy

So what do the bottom two boxes in *Figure 10.8* really mean?

Our technology value-add will likely remain what it is today. That is your current product sector expertise. Added to it is your particular slant on service value-add that can optimize consumption. For Apple, it has at least three components: their unique user interface and designs, their in-store support capabilities, and their content services like iTunes and the App Store. When they think about their products, they don't consider these service value-adds as an afterthought once the core products are complete; they think of all the components together at the design stage. This is what we need to do. Another thing they do is to consider the whole ecosystem and how they can control it. This includes third-party content providers like app developers and entertainment companies. They aren't just trying to change how products work—they are trying to change how markets work.

For enterprise tech companies, the second layer of the Consumption Model is another big mental shift. If we are really executing well at the core levels in the previous paragraph, then we have effectively diversified our value-add from being just about a piece of technology to a much broader set of solutions. Sure, customers can still buy our basic product plus basic service, and some will still want to do that. But that is not where the growth will come from and not where consumption will be optimized. We need a much broader range of solution types, and some of them may seem very awkward. Obviously we will have our cloud-based offers that we have talked about throughout the book. And we all see the continued growth of managed service and outsourcing solutions. They are hot and will get hotter. But maybe a customer who is using another company's products today just wants to contract for your value-added Consumption Service expertise. Maybe that initial contract has none of your products in it. That will seem unacceptable to our product playbook DNA. But in the new normal, that is the path to lots of product business—it might not be until year two, but it will ultimately be there.

We also may need to consider very different solution pricing models. We will probably need to play not only in pay-per-use models like these, but also dabble in business outcomes. Maybe we will share in total customer cost reductions. Maybe we will share in new revenue growth. These are different kinds of offers that require different kinds of corporate DNA.

These new solution types are essential because of the real changes characterized by the middle set of boxes in *Figure 10.8*. In many of these new solution types there will be no revenue at the contract signing and no maintenance contract to move them onto. So the customer's ability to consume becomes the key—the most important single consideration from the time we start the sales process until the time we go to our graves. How do we do that? Well, we have to design the optimum technology solution for their business needs, and we have to optimize the customer's ability to consume. These are both variations on design services, but one's goal is to design the technology, and the other's is to design the customer's business processes and end-user skills. These capabilities are high value and in short supply industry-wide. We are going to need far more of these kinds of people, and they will not just sit in service roles—they also will be the core expertise of our sales motions.

At the top level are all the consumption-based functions that we've talked about throughout the book. Here sits all the internal organizations that interact with existing customers and end users as they consume our solutions. This includes Consumption Development, who is listening in real time to customer needs and developing new functionality. Consumption Marketing is using our products' new Consumption Innovation and In-Product Upsell capabilities to drive more feature adoption and MT purchases. Our Account Services Organization is not only supporting the customer and the end users, but they are also optimizing billable consumption 24 × 7 × 365. They are working with Consumption Professional Services to address the high-level roadblocks that may impede growth, like security or compliance concerns and business process bottlenecks. They also are working with the sales function, where they see big incremental opportunities at big customer sites.

This is a full-blown Consumption Model in action.

It doesn't mean we don't need to innovate anymore. It means that we need to innovate around a different objective: consumption of value. Doing that well will mean diversifying your DNA to include product innovation, service innovation, and business model innovation—including working with other companies' products.

There are lots of changes coming to the tech industry:

- There could be lower margins across the board. Higher volume and new directly served markets like SMB will be key. Remember: Complexity is the enemy of volume.
- Average transaction prices will continue to fall.

- Account profitability will be far longer in coming, but customer lifetime value might be higher.
- Traditional strangleholds that kept even marginal customers locked in and paying maintenance will begin to dissolve. Maintenance could dry up.
- Subscription-based pricing models will change cash flow and revenue recognition patterns.
- Established cash reserves will need to fund account transitions to consumption-based purchase models.
- We will experiment far more with value-based or outcome-based pricing.
- Contracted monthly recurring revenue (CMRR) will become the hot, new key performance indicator.
- Existing long-term contracts and SLAs will get restructured and renegotiated.
- Churn will take on a whole new importance, just like it does for cellular carriers, ISPs, and content companies.
- Customers will be seen as assets to manage. Not all will provide the same level of return based on their internal constraints to consumption. Cost per managed asset (CPMA) will become a key sales concept and will become a qualifying factor in new customer investment decisions.

The end result is this: How tech companies operate and how they make money will change more in the next 10 years than they have in the last 40 years. These changes will mean that lots of the common wisdoms, experiences, skills, and models that have gotten us this far may no longer apply. The changes are major, and the courage to make them will not be easy to come by. There will be a lot of entrenched forces telling you to hang on to the past. We think that your leading-edge customers, your next generation of internal leaders, and your long-term shareholders are of a different mindset.

Consumption Economics means that there are new rules of tech. Those companies that get them right will rule.

11 The "S" Stands for Services

LET US BRIEFLY TELL YOU WHY TSIA IS WRITING A BOOK ABOUT the future of tech.

It's simple. The future of tech is about the evolution of our industry toward being offered as a service. These services will not only be easier for customers to adopt but also easier for them to abandon. Gone will be many of the mysteries and costs of complexity, the curtain that underperformance hides behind.

If your doctor does not make you well, you get a second opinion. If employees hate their company's web-conferencing service, either IT finds them an alternative or they do so themselves. If your laptop is unreliable, you get a new one from a company that takes ease-of-use seriously. This is the way life in tech is heading. Admittedly, it is starting at the bottom of the complexity stack, especially in the enterprise. But how much more evidence do we need that the cloud is *The Natural*?

Along with hundreds of global companies that make up the Technology Services Industry Association community, we carefully focus on creating service best practices and IP. TSIA is a place where technology executives from different companies can work together with the common aim of learning how to better manage and optimize their own service businesses. That is about to become not just a question for their own organizations. It is about to become the whole company's question.

As we look ahead, we see a role for service organizations in the cloud that is far more strategic and exciting than at any other time in our industry's history. We see an evolution in customer service, professional services, education services, and operate services that could change the world through better consumption of technology's value. It is our goal to make a lasting contribution to this huge and important transition. But as we all know, this transformation is not limited to traditional service organizations—every corporate function will need to be rethought in the age of the cloud. So the TSIA community is also at work to bring the lessons of the "________ as a service" business model to product-focused sales, development, and marketing organizations.

Consumption Economics is coming. It demands a new business model for the technology industry. We don't see any company fully executing it, but we do see many on the way. TSIA and the leading companies we work with are on the brink of some real breakthroughs.

The "S" in TSIA stands for services. The cloud is all about services. Let's go.

Endnotes

Chapter 1

1. Source: 2010 Software Success Survey.
2. Source: TSIA, Neochange, Sand Hill Group 2009 research.
3. Source: Accenture, 2008.
4. Source: Wall Street Journal, January 3, 2011.
5. Visit www.wordbanter.com to see the thread, http://www.wordbanter.com/showthread.php?t=63381.

Chapter 2

1. For a great video of Larry Ellison questioning the hype of cloud computing, visit http://www.youtube.com/watch?v=UOEFXaWHppE.
2. To see a video of Jeff Bezos' keynote, visit http://mitworld.mit.edu/video/417.

Chapter 4

1. Hawkins, Wendy. April 8, 2010. *Game Mechanics: What Businesses Can Learn from Gaming.* http://www.springbox.com/insight/post/Game-Mechanics-What-Businesses-Can-Learn-From-Gaming.aspx.

Chapter 6

1. Yeeguy. January 17, 2011. "How *Facebook* Ships Code." *FrameThink* blog. http://framethink.wordpress.com/2011/01/17/how-facebook-ships-code/.

Chapter 7

1. Dreze, X., S.J. Hoch, and M.E. Purk. 1994. "Shelf Management and Space Elasticity." *Journal of Retailing* 70:301-326.
2. Ciuti, Luigi, personal communication. February 21, 2005. Also see: Sigurdsson, Valdimar, Hugi Saevarsson, and Gordon Foxall. *Brand Placement and Consumer Choice: An In-Store Experiment.* http://www.ncbi.nlm.nih.gov/pmc/articles/PMC2741065/#jaba-42-03-26-Dreze1.

Chapter 8

1. Fryer, Bronwyn. May 7, 2001. *Siebel: New Ideas, Old Economy.* http://hbswk.hbs.edu/archive/2226.html.

Chapter 9

1. Source: TSIA tracks 50 of the largest providers of technology solutions and reports on key trends such as product-service mix.
2. Includes enterprise IT, consumer technology, medical devices and healthcare IT, and industrial automation.

Chapter 10

1. *The Natural* is a 1984 movie starring Robert Redford, Robert Duvall, and Glenn Close. An average baseball player comes out of seemingly nowhere to become a legendary player with almost divine talent.

Index

C